Invisible Colleges

Invisible Colleges

Diffusion of Knowledge
in Scientific Communities

Diana Crane

The University of Chicago Press · Chicago & London

The University of Chicago Press, Chicago 60637
The University of Chicago Press, Ltd., London
© 1972 by The University of Chicago
All rights reserved. Published 1972. Second Impression, 1975
Printed in the United States of America
International Standard Book Number: 0-226-11857-6
Library of Congress Catalog Card Number: 72-182088

To My Father

Contents

2. How do S C's evolve?
On these definite stages?

3.

Acknowledgments

A NUMBER OF PEOPLE HAVE CONTRIBUTED TO THIS BOOK IN various ways. Derek J. de Solla Price was helpful not only through his writings, which stimulated my interest in this subject and contributed to my understanding of it, but also through his interest in my research on the sociology of science and by generously making research funds available. The encouragement and advice of Ivan Vallier were of enormous aid. Michel Hervé has read numerous drafts of this book and contributed substantially to my thinking. The comments of my students and colleagues at Yale University and at The Johns Hopkins University have also been valuable. I am grateful to Walter Wallace for a critical review of an earlier version of the manuscript.

As in most sociological studies, I am much indebted to my respondents who gave their time and experience to this study. I am grateful to Everett Rogers for permitting me to analyze materials from the Diffusion Documents Center at Michigan State University. Daniel Gorenstein and Robert Chipkin were very helpful in the initial stages of my study of the mathematicians who are concerned with the mathematical theory of finite groups. Daniel Gorenstein also read and commented upon an earlier version of the manuscript. David Vachon, Edwin Olson, and Erik Steiner developed the computer program that was used to trace sociometric connectedness from an earlier version published by Coleman (1964).

Mrs. Jean Castells and Mrs. Barbara Florian were fine research assistants, and Mrs. Florian and Mrs. Elizabeth Yager patiently typed various drafts of the manuscript.

This research was supported by Grants GN-527 and GS-2205 from the National Science Foundation.

Portions of the material for this book have appeared in a somewhat different form in my articles in the following publications: "Fashion in Science: Does It Exist?" *Social Problems* 16 (1969): 433–41; "Social Structure in a Group of Scientists: A Test of the 'Invisible College' Hypothesis." *American Sociological Review* 34 (1969): 335–52; "La diffusion des innovations scientifiques." *Revue Française de Sociologie* 10 (1969): 166–85; "The Nature of Scientific Comunication and Influence: A Theoretical Model." *International Social Science Journal* 22 (1970): 28–41.

1
Introduction

IN THE LAST TWO DECADES, DRAMATIC INCREASES IN THE scope and volume of scientific research have occurred, as may be illustrated by the fact that the amount of scientific literature is doubling approximately every ten years (Price 1963). For the scientist who needs to locate particular items of scientific information and for the documentation specialist who must make them readily available, the organization and management of this huge and expanding store of information is a serious problem. Increasingly radical solutions are being proposed. For example, some experts would like to scrap scientific journals and distribute their contents piecemeal. Information retrieval and delivery systems are being developed to enable scientists to locate information quickly and effectively.

But in sharp contrast with the attention being paid to how knowledge is stored, distributed, and used, relatively little attention has been paid to why and how knowledge grows. In seeking to describe the informal but specialized social system that produces basic scientific knowledge, this book treats one aspect of this problem.[1]

The growth of scientific knowledge, like that of most

1. As opposed to applied science, basic science is concerned with problems whose solutions are not perceived as having practical applications.

natural phenomena, takes the form of the logistic curve. The logistic curve has been fitted to the cumulative numbers of new publications appearing per year in scientific disciplines (Price 1961, 1963). This means that the growth of numbers of new publications is passing through the following series of stages: (1) a preliminary period of growth in which the absolute increments are small although the rate of increase is large but steadily decreases; (2) a period of exponential growth when the number of publications in a field doubles at regular intervals as a result of a constant rate of growth that produces increasing amounts of absolute growth; (3) a period when the rate of growth declines but the annual increments remain approximately constant; and (4) a final period when both the rate of increase and the absolute increase decline and eventually approach zero. Price has argued that basic science is currently in the second phase of growth but that, as a result of shortages of resources and manpower, it will eventually enter the third and fourth stages.

Why should the growth of scientific knowledge take this form? Price does not offer an explanation. The thesis that will be presented in this book is that the logistic growth of scientific knowledge is the result of the exploitation of intellectual innovations by a particular type of social community. The question to be examined is how scientific communities affect the growth of knowledge.

It is not surprising that the sociological analysis of the production of science, ideology, philosophy, religion, literature, and the arts has been largely neglected since few sociological problems are so complex as that of understanding the social institutions that produce ideas. In dealing with these types of phenomena, the sociologist is faced with the problem not only of understanding the social relationships between individuals but also of understanding the ideas themselves, which can be highly technical and abstruse. Even if the sociologist elects not to become an expert on the details of his subjects' intellectual productions, he cannot ignore the nature of these activities entirely since presumably they affect in some way the social relationships among his subjects, and the latter in turn affect the production of ideas.

The social institutions that produce scientific ideas provide an excellent example of the kinds of difficulties that the sociologist faces in conducting such studies. For example, in addition to a relatively small number of professional associations, there are large numbers of scientific communities that form around the study of particular scientific questions. It is theoretically possible (although the opposite will be argued in these pages) that the diversity in the types of scientific communities is as great as the diversity among types of scientific problems. As will be shown in this book, these scientific communities are both distinct entities and interconnected with one another in ways that are as yet only vaguely understood. Understanding how the scientist pursues his task of creating new knowledge that will become part of this complex structure can require the contributions of several disciplines, including philosophy, history, psychology, and sociology.

There is currently renewed interest among sociologists in studying all of the social institutions which produce ideas (Albrecht et al. 1970; Curtis and Petras 1970). At the same time, among sociologists of science, there is emerging considerable controversy about how and on what levels the sociological investigation of science should be pursued. In this chapter, I will discuss various approaches to the sociological study of science and will define the particular approach to be used in this book.

Approaches to the Sociology of Science
The sociological study of science is a relatively recent development. Although the first studies appeared more than thirty years ago, it is only in the last decade that a substantial number of publications have appeared. Several orientations toward the field can be identified at the present time.[2] The first approach is concerned with investigating the interrelationships between science and other social institutions.

2. This brief review excludes several other aspects of the sociological investigation of science, such as studies of scientific creativity, science policy, and research organizations, which are for the most part unrelated to the problems being discussed.

How have other institutions influenced the origins and development of science? The reciprocal question, how has science influenced these other social institutions, has received much less attention from sociologists.

The second approach is concerned with the study of science as a social system. Science is treated not as a body of knowledge or as a set of methods and techniques but as:

. . . the organized social activity of men and women who are concerned with extending man's body of empirical knowledge through the uses of these techniques. The relationships among these people, guided by a set of shared norms, constitute the social characteristics of science (Storer 1966:3).

Two other approaches have been used by members of other disciplines. One is identified with a few historians of science whose work has had considerable influence upon sociological thinking about science (Holton 1962; Kuhn 1962; Toulmin 1963). They are concerned with identifying the characteristics of scientific knowledge that bring about change and development in scientific ideas rather than with identifying external influences upon this process. Another approach that has been utilized primarily by information scientists examines the way scientists use the scientific literature, particularly the extent to which they cite recent literature compared to older sources, literature in their own fields compared to other fields.

The first approach stems from an earlier tradition, the sociology of knowledge that was concerned with the influence of social institutions upon the development of ideas. For example, the basic theme of Marx's contributions to this tradition was that social relationships based upon the economic system of a society influence the system of ideas in that society. A man's position in the social class structure affects the ideas he accepts and produces. Weber took the opposite position. In his view, ideas are not entirely affected by material factors; they also play an independent role in society. He examined the influence of religious values upon the development of the economy as well as the influence of ideas upon the relationships between social classes. Merton's early study (1957) of the origins of scientific activity in Western society was an attempt to extend Weber's ideas. Studies by Merton (1957) and Barber (1962) of the relationships between science and the political system of a society also belong to this tradition.

Another theme in the sociology of knowledge is exemplified by Mannheim's hypothesis that there is a relationship between the particular structure and goals of social groups and the types of general world-views that they create and accept. For example, conservative groups create and maintain a static view of the world, whereas progressive or revolutionary groups adopt a dynamic view. He was also interested in the fact that members of different generations have different views of their experiences and purposes and that these sometimes result in conflict between generations.

The sociology of knowledge has provided an orientation for the sociology of science, but its theoretical insights have not (with the possible exception of Weber's work) exerted a strong influence. The kinds of theoretical issues raised by sociologists of knowledge have received little attention in recent years. The basic theme of the sociology of knowledge, that the content of ideas is in some way influenced by social structure, is almost entirely ignored. Part of the explanation for this neglect lies in the vagueness of these ideas and in the difficulties involved in testing them empirically. For example, as Merton has shown in a critique of Mannheim's work (1957), Mannheim's failure to specify the way in which knowledge is related to social structure has made it difficult to develop empirical studies based on his ideas. If the social structure influences the development of ideas, how does this influence occur?

Among sociologists, concern with the relationship between science and society is most alive in the writings of Joseph Ben-David, who has investigated the organizational prerequisites for the development of science in western Europe (Ben-David 1964) and for the emergence of new scientific disciplines (Ben-David 1960; Ben-David and Collins 1966). At the present time, much of the writing on the relationships between science and society is being done by political scientists who are concerned primarily with the relationships between scientific institutions and national governments, and by natural and social scientists who are seeking to influence national science policy. The amount of interaction between these groups and sociologists who study science is small.

In recent years, the major interest of sociologists of science has been in the second approach, the study of the social system of science. The earliest work on the social organization of science constituted a definition of the scientific ethos. Five principal norms governing scientific activity were postulated (Merton 1957; Barber 1962): (1) the scientist is expected to evaluate new knowledge critically and objectively; (2) he is expected to use his findings in a disinterested fashion; (3) scientific merit should be evaluated independently from the personal or social qualities of the individual scientist; (4) the scientist does not own his findings; secrecy is forbidden; (5) he is expected to maintain an attitude of emotional neutrality toward his work.

These writings presented an idealized conception of science. The scientific ethos is supported by scientists not only because it is technically efficient to do so but because these values "are believed right and good." The element of motivation was largely ignored; the level of deviance was considered to be very low. Disputes over priority in scientific discoveries were given considerable attention but were interpreted using the same framework. That is, such disputes were seen as a consequence of adherence to, rather than deviance from, the norms of science that place a high emphasis upon both originality and criticism.

Hagstrom (1965), however, recognized the existence of deviance from scientific norms and became concerned with the conditions that produce conformity to or deviance from these norms and values. In his view, scientific behavior is an exchange of information for recognition. Information is a gift that the scientist gives to other scientists in return for which he expects recognition. In making the gift, the scientist conforms to scientific standards concerning the production of scientific work and does not act solely from his desire for recognition. If the scientist is to receive recognition for his work, he must produce answers to questions that are of interest to other scientists. This sociological interpretation of the behavior of individual scientists led Hagstrom to consider the nature of scientific disciplines in order to understand how scientists select problems for study.

He argues that the major source of deviant behavior in science is competition for recognition, and that it is possible to differentiate between disciplines in terms of the likelihood that competition will result: "Competition results when scientists agree on the relative importance of scientific problems and when many of them are able to solve these problems" (Hagstrom 1965: 73). Specifically, competition increases in a scientific field when (a) agreement about the relative importance of scientific problems increases; (b) the number of specialists able to attack the problems increases; (c) the degree of precision obtained in particular research results increases—so that it is clear whether or not a problem has been solved. The specialty is a kind of environment—an institutional context influencing the way the scientist behaves. In order to understand scientific behavior, some knowledge of the state of the specialty is needed.

Hagstrom's concern with the differences between scientific disciplines reflects the influence of the third orientation toward the sociological aspects of science. These historians of science are interested in the way scientific knowledge grows and cumulates. For example, Kuhn (1962) has proposed a model of scientific change that has stirred considerable controversy. Briefly, he argues that growth and development of scientific knowledge occur as a result of the development of a paradigm or model of scientific achievement that sets guidelines for research. After a period of "normal science" during which the implications of the paradigm are explored, facts that the paradigm cannot explain become inescapable. The field then goes through a period of "crisis" during which a new paradigm is proposed and eventually accepted.

Finally, analyses of the structure of scientific literature have shown that each new piece of scientific literature is related to several previous pieces that in turn are related to earlier work (Price 1965a). The scientific literature grows very rapidly. The "life" of a paper is very short, with the exception of a few classics. Papers published five years ago are "old." Papers published more than fifteen years ago are almost useless in many scientific fields (Burton and Kebler 1960); the research "front" has moved too far beyond them.

Currently these approaches to the sociology of science are being reevaluated. For example, a group of British sociologists has been critical of the normative approach on the grounds that there is little empirical evidence to support the existence of the scientific ethos and that the actual behavior of scientists, in contrast to what they say they do, shows examples of deviance rather than conformity to these norms (Barnes and Dolby 1970; Mulkay 1969). They are also critical of this approach because it excludes an examination of the cognitive aspects of science. Instead of one scientific ethos influencing the entire scientific community, they prefer Kuhn's theory (1962) that scientists in each field develop a consensus concerning the important problems that provides guidelines for further research. While recognizing the difficulties with Kuhn's work (Martins 1970), they argue that sociologists should undertake the study of the cognitive cultures of scientific fields.

As some of these writers show (King 1970; Martins 1970), each of the three approaches to the sociology of science implies different assumptions about the nature of scientific proof and the process of scientific growth. The normative approach implies a positivist view of the nature of scientific proof. Scientists collect evidence to test hypotheses and theories and agree unequivocally about their findings. This view, which Hagstrom calls "the positivist ideal," considers controversy, or the existence of two opposing and irreconcilable interpretations of the same empirical findings, as unjustified. A recent study using the normative approach (S. Cole 1970) concluded that resistance to scientific discovery is rare in contemporary science and that the social structure in which scientific knowledge is pursued has little effect upon the development of scientific ideas.

Since controversy is contrary to the norms of scientific behavior, most scientists adhere to the positivist ideal that hypotheses and theories are susceptible to unambiguous verification (Hagstrom 1965: 255–59). As Hagstrom has shown, however, this ideal is applicable only under certain conditions that do not exist in all research areas. Hypotheses and theories are likely to be susceptible to unambiguous empirical verification when they are sharply restricted in scope, when highly precise quantitative data is obtainable, and when alternative hypotheses do not have far-reaching implications for the field as a whole.

Underlying both the earlier sociology of knowledge tradition and recent work by historians of science is the view that the growth of scientific knowledge is not an entirely rational, logical, continuous process. Instead, Kuhn (1962) describes periods of "normal science," when scientific growth does conform to the positivist ideal, punctuated by periods of discontinuity and crisis when opposing theories are the subject of long and difficult controversies among scientists. King compares the scientist to a legal judge:

. . . a man engaged in the interpretation, elaboration, modification and even on occasions overthrow of a professional tradition of practice rather than an automaton whose activities are finally monitored by a fixed, inexorable logic (1970: 46).

Unfortunately, a major dilemma that sociologists (and others) face in studying science is that any approach that they take to the subject has epistemological implications. If the sociologist refuses to consider the content of scientific ideas on the grounds that it is not affected by sociological considerations, he is implicitly accepting a theory of knowledge in which the scientist's decisions about his work are based entirely upon logical considerations. On the other hand, the proposal that social factors affect the development of scientific ideas leads to an epistemological paradox: if the sociologist argues that scientists cannot be entirely objective in their approach to science, then his own analysis of science must be subject to the same flaw. This issue was raised in connection with the work of Mannheim and, although Mannheim excluded from his consideration the natural sciences and devoted his attention largely to the study of political ideologies, concern with this problem inhibited the development of his ideas and also the subsequent impact of his work.

Several ways of dealing with this paradox have been suggested. For example, King (1970: 59) recommends a period of "epistemological agnosticism" during which the demands of epistemology would simply be ignored in favor of studying social influences upon the development of ideas. De Gré (1970) also argues that the sociology of knowledge need not concern itself with epistemological problems. According to him, the sociology of knowledge is concerned with the meanings of ideas to their proponents and in relating these ideas to the sociohistorical conditions within which they occur, not with the relationships between these ideas and the actual facts. The sociologist of knowledge is not concerned with the logical validity or empirical truth of the ideas he studies. Ignoring the implications of a sociological interpretation of scientific behavior for the objectivity of scientific behavior, however, does not resolve the epistemological difficulty.

Martins (1970) doubts that the epistemological problems of "cognitive sociology," which is the name he prefers for the sociology of knowledge, can ever be entirely resolved. Some aspects of "cognitive sociology" can escape them, but he doubts that the subject can be "permanently vaccinated" against these difficulties. He argues that the epistemological issues must be faced and are not a sufficient reason for dismissing the sociology of knowledge altogether.

While it is possible for sociologists to accept the positivistic interpretation of scientific change and to concentrate their attention upon questions concerning conformity and deviance to scientific norms, the exchange of recognition, and the structure of scientific organizations, sociologists who take a nonpositivist position on epistemological questions argue that some of the most important questions concerning scientific institutions, those that concern changes in the development of scientific ideas, are being ignored.

In spite of their apparent conflict, what seems to be taking place is the beginning of a synthesis of these three approaches to the sociology of science. In the process of undergoing such a reconciliation, each will be altered considerably. For example, the sociology of knowledge was primarily concerned with the way in which ideas were influenced by the position of intellectuals in the social structure of their society. The social structure of intellectual communities was largely neglected. While the normative approach has been concerned with understanding the values and norms of science, they have tended to view the scientific community as a single entity and have postulated very general norms without considering the variations that may occur in different specialties or subspecialties and in different time periods. What is needed is a merger of these two traditions in an investigation of how the social systems of scientific communities influence the development of scientific ideas, using insights provided by the third approach regarding variations in the cognitive structures of scientific specialties and by the fourth approach concerning the nature of scientific literature. Appropriately enough, one of the leading members of the third approach is also becoming more concerned with the relationship between scientific communities and the process of scientific change (Kuhn 1970).

Orientation and Research Design for a Study of Scientific Communities

How to explain change and development in scientific thought is a fundamental issue in the study of scientific institutions, since the primary goal of these institutions is to produce new knowledge. One of the first sociological questions that must be asked in this regard is whether scientific communities and variations in communication patterns among scientists actually affect the development of knowledge. If such an effect can be shown by examining the effect of scientific communities upon the accumulation and acceptance of ideas, it would then be appropriate to examine in greater depth the cognitive cultures of such communities in order to specify more precisely the interaction between the cognitive and social components of science.

If the effect of scientific communities upon the development of knowledge is to be examined, it is necessary to find a means of identifying these communities. Studies of citations (Price 1965a) show that about half the references in each group of new scientific papers link them to a small group of earlier publications, most of which are close to them in time. The other half of the references link the papers apparently randomly to a very sizable part of the scientific literature. These findings indicate that the literature of basic science consists of tightly knit clusters of papers, each of which is loosely linked to a large number of other clusters. The clusters represent research areas, sets of closely related problems that, as will be shown in subsequent chapters, are viewed by the scientists who study them as discrete entities.

Examination of the cumulative numbers of new publications and of new authors publishing for the first time in research areas shows that the growth of research areas fits the same logistic curve that Price (1963) has used in describing the growth of new publications in scientific disciplines to which research areas belong.[3] After a relatively slow start, both disciplines and research areas exhibit a period of exponential growth that is followed by a period of linear growth and finally by a period of slow and irregular growth. Science as a whole appears to consist of hundreds of research areas that are constantly being formed and progressing through these stages of growth before tapering off.

Price (1961) has estimated that the normal rate of exponential growth in science is such that the scientific literature doubles approximately every ten years. The rate of growth of research areas within science, and even certain

3. For logistic curves illustrating the growth of research areas see figures 2–6 and 9–12. Contrary to the practice that is generally followed in the presentation of logistic curves, these figures show cumulative increases by year on an absolute scale rather than on a logarithmic scale. Plotting such figures logarithmically approximates a straight line that indicates the presence of exponential growth. The central argument being presented here stresses the differences in the characteristics of various stages of the logistic growth of research areas; for this reason, it seemed preferable to present the figures in such a way that these four stages are clearly visible to the reader.

disciplines, appears to be more rapid.[4] This can be explained by the fact that science as a whole consists of research areas at various stages of growth, both exponential and nonexponential. Some fields remain relatively dormant for long periods of time. Either they have never attracted widespread interest or they have done so in the past and have entered a period of prolonged decline. As a result, the rate of exponential growth for science as a whole is slower than that of some individual areas.

No research area is completely isolated from other areas. Social and ideational links hold the various segments of knowledge together and permit the diffusion of ideas from one area to another, but in ways so complex that it is difficult to identify unequivocally a particular research area. Even the labels that scientists use to describe their research problems are constantly changing. For example, physicists who were asked to describe the major subfields of their discipline indicated that they viewed their discipline as "a fluid subject, with shifting boundaries, and eruption and subsidence of topics within these boundaries" (Anthony et al. 1969: 731).

This suggests that the term that best describes the social organization of the entire set of members of a research area is the concept of the "social circle" (Kadushin 1966, 1968). The exact boundaries of a social circle are difficult to define. The boundaries of this group in terms of its total membership are also difficult to locate. Each member of a social circle is usually aware of some but not all other members. The members of a research area are geographically separated to such an extent that face-to-face contact never occurs between all members and occurs only periodically among some. Indirect interaction, interaction mediated through intervening parties, is an important aspect of the social circle. It is not necessary to know a particular member of a social circle in order to be influenced by him. Not only can a scientist be influenced by publications written by authors whom he has

4. The literature of physics is now doubling every eight years (Anthony et al. 1969: 713). The literature of sociology doubled every three years from 1954 to 1965 (see fig. 15). See Table 28 for the rates of exponential growth in each of the areas shown in figures 2–6 and 9–12.

never met, but he can also receive information second-hand through conversation or correspondence with third parties. There is no formal leadership in a social circle although there are usually central figures. Authority relationships are contrary to the professional norms that underlie scientific activity (Hagstrom 1965). Scientists are supposed to advise and criticize but not to command each other.[5] The social circle is not well instituted compared to the bureaucracy or even to less formalized entities such as the tribe or the family. Members of a social circle come together on the basis of their interests more often than on the basis of propinquity or ascribed status. Members of a research area are brought together by their commitment to a particular approach toward a set of problems.

The amorphous character of research areas complicates the problem of defining the membership of the social circles that study them. In this study, comprehensive bibliographies were used to define the membership in two research areas, one in rural sociology and one in mathematics.[6] In each area, the bibliography was compiled by one of its most active members. Use of such a bibliography entailed accepting the judgment of its compiler regarding the scope of the field and its

5. There are situations where scientists have authority over other persons, but the latter are generally not professionals engaged in basic research. They are more likely to be graduate students, who are not considered full-fledged scientists, or research assistants, who usually do not have advanced degrees.

6. The names of the rural sociologists were obtained from Rogers (1966). The mathematicians' names were obtained from Gorenstein (1968). A graduate student specializing in finite groups searched *Mathematical Reviews* from 1948 to mid-1968 in order to locate additional authors. Sixty-one authors were obtained from Gorenstein's bibliography and forty-one from *Mathematical Reviews*.

An alternative method of defining a research area would have been to locate a few publications identified with the area in question and to trace publications that cited them over a period of time. In turn, publications citing these publications could also be used. Identification of the starting points would be crucial to this method, which is also tedious since the *Science Citation Index* that shows these connections was not published before 1961. In addition, a certain number of publications cited would belong to other areas and identifying these would be problematical.

relevant publications. Since scientists themselves do not always agree about the classification of specific pieces of research in terms of either the labels they are currently using or the categories found in indexing and abstracting services, it seems likely that no two bibliographies compiled by active members of an area would be identical. Disagreement would presumably be highest concerning pieces of research that are peripheral to the area and could be considered to belong primarily to another research area by at least some members. Utilizing bibliographies compiled with a view toward completeness by knowledgeable members of research areas appeared to be the simplest way to identify both the intellectual and the social components of the area.

The first area that was studied was part of the research specialty, rural sociology, and was concerned with the study of the diffusion of agricultural innovations. A research area in sociology was selected so that the investigator would be familiar with the nature of the research and with the names of prominent researchers both within and outside the area. Since hypotheses about communication networks in science were developed from observations of the behavior of scientists in fast-moving specialties within physics (Price 1963), the selection of a research area in the social sciences, which are often thought to grow more slowly and less efficiently than the natural sciences, might seem inappropriate. All the same, an analysis of 403 papers published in rural sociology between 1941 and mid-1966 revealed that this area possessed several of the characteristics that have been found in the literature of the natural sciences (Price 1961): (1) growth of the field (number of papers published per year) had progressed through the first three of the four stages that Price has described as being characteristic of scientific literature; (2) the number of new authors entering the field each year showed the same series of stages of growth; and (3) a few authors in the area had been highly productive and a majority had produced one or two papers. Although the field contained a larger proportion of single-author publications (61 percent) than would be expected for a research

/.

area in the natural sciences,[7] so many factors are believed to influence the amount of collaboration in a research area, and these factors are so imperfectly understood, that this characteristic did not seem important enough to disqualify the area for a study of this type.

A second aspect of the research area that might be considered to limit its usefulness for a case study of this sort is the type of situation in which research in this field has been conducted. For reasons related to the history of rural sociology, research in the diffusion area has often been financed and conducted in agricultural experiment stations (Kaufman 1956). As a result, some of the research has had an applied character. This research, however, is not carried out under the conditions of restricted communication that usually characterize applied research in industry. Rural sociologists studying diffusion have been in excellent communication with one another (Rogers 1962: 38). Thus, the applied character of some of the research does not appear to have inhibited the development of informal social organization in the field.

The second field selected for this study was a research area in algebra that was concerned with the theory of finite groups. In this case, an effort was made to select an area from a discipline in which the character of research is very different. Data from studies conducted by other investigators of research areas in the biological sciences, the physical sciences, and psychology (Crawford 1970; Gaston 1969; Libbey and Zaltman 1967; Mullins 1968a, b, c; Project on Scientific Information Exchange in Psychology 1969; Zaltman 1968) will be reviewed for comparative purposes.

In the mathematics area also, analysis of the 305 papers published between 1906 and 1968 showed that the growth of the field in terms of cumulative numbers of new authors and new publications had also progressed through the first three stages of the logistic curve. The proportion of single-

7. Estimates of the proportion of single-author papers range from 32 percent for chemistry in 1963, 23 percent for biomedical literature in 1963, and 17 percent for physics, chemistry, and biological sciences combined between 1950 and 1959 (Clarke 1964; Zuckerman 1967).

author publications was very high (86 percent), but this is typical of mathematics as a discipline, as has been shown by Walum's study (1963: 64–66) of the mathematical literature.

In both research areas, a member was defined as someone who had authored or co-authored at least one paper appearing in the comprehensive bibliographies. There were 221 rural sociology authors and 102 mathematics authors. In each research area, respondents were sent lists of their publications in the area as listed in the source bibliographies and were asked to reply to a questionnaire in terms of their research in the area as represented by those publications only.[8] This was to avoid responses that described the respondent's associations with researchers in other research areas where he might have been active. In the rural sociology area, publications included theses, papers presented at professional meetings, and agricultural experiment station bulletins and reports, as well as journal articles and two books. In the mathematics area, the publications consisted of journal articles only. These definitions reflected the nature of publications in these areas.

Both bibliographies covered lengthy time periods. The listing for the rural sociology area spanned the period 1941–66. The listing for the mathematics area covered the period 1907–68. In order to assess the nature of the social relationships in these areas and the manner in which they had changed over time, it was essential to obtain information from individuals who had published in these fields at different times. This meant that some respondents were replying about events that had taken place many years ago. It was considered preferable to have such information in spite of possible

8. A sample questionnaire is reprinted in the appendix. Questionnaires were sent to all individuals listed in the bibliographies (Rogers 1966; Gorenstein 1968; *Mathematical Reviews*—see note 6) for whom addresses could be obtained. Questionnaires were returned by 67 percent of the rural sociologists and by 63 percent of the mathematicians. The response rates were higher when those for whom addresses were unavailable were excluded: 79 percent and 74 percent, respectively. In both samples, American and Canadian respondents who did not return the questionnaire were interviewed by telephone.

unreliabilities inherent in retrospective replies. Since parallel studies of the literature were being conducted in each of these areas, the latter provided a check on the findings from the questionnaires.

Two types of data were used in the studies of the literatures of these areas. The author of the comprehensive bibliography in the rural sociology area (Rogers 1966) had supervised a content analysis of all publications in his bibliography that contained empirical data. For each publication, all dependent and independent variables utilized in the research reported in these publications were coded and punched onto IBM cards (Rogers et al. 1967). These data, which provided an unusual opportunity to trace the dissemination of scientific ideas, were recoded and used in making the analysis presented in Chapter 4.

The first use of a dependent or independent variable in a publication in a research area represents an innovation in the form of a new hypothesis or a revision of a previous hypothesis. This definition of an innovation was used. In the study of the literature in the rural sociology area, each variable or "innovation" was taken as a unit of analysis. All publications using that variable were treated as sources of information concerning the diffusion of that variable from one author to another in the research area.[9] When an innovation appeared for the first time in two or more publications in the same year (suggesting simultaneous independent innovation), such publications were assigned to different categories on a fractional basis. The innovations ranged from relatively minor variations on previous innovations to the introduction of new and quite different concepts.[10]

9. The available data showed which scientists used the variables and the dates when they used them in the literature of the research area. The process of diffusion can be traced in this manner, but whether variables were "reinvented" by later users could not be ascertained from this data. In the questionnaire sent by the author to members of the research area, they were asked to name persons or publications that had influenced their selection of problems. Using this information, it was possible to ascertain how frequently each of the innovators had been named.

10. When "major" and "minor" innovations were distinguished in a preliminary analysis, the results were similar to those reported in chap. 4.

In the mathematics area, it was not possible to obtain information about the introduction of specific innovations. Instead, the frequency with which a paper was cited was used as a measure of its innovativeness. Within a research area, frequent citation indicates that a paper contains information that has been useful to other members. All references by papers in the mathematics area were obtained from their bibliographies. It was then possible to categorize a paper in terms of the characteristics of its senior author, the number of senior authors in the area who cited it, the productivity of the citing authors, the sizes of the groups of collaborators citing it, and the timing of the citations.[11]

As this description of the study shows, the simplest and most objective indicators of the intellectual and social aspects of these research areas were utilized. Growth in knowledge was measured using the number of publications appearing per year throughout the entire history of each research area. Relationships between ideas were measured using references appearing in publications. Relationships between scientists were measured using the extent to which they named each other when responding to questions concerning their informal communication practices or sources of important influences upon their work.

Several writers have argued that growth in gross numbers of publications is not a good indicator of the growth of knowledge on the grounds that only a few publications are heavily utilized in later scientific work and most are seldom referred to in later publications. Weinstein, for example, has argued on this basis that counting the number of publications produced in an area is a misleading estimate of its growth. J. Cole (1970) makes a similar argument and suggests that the seldom-cited publications are unnecessary for the development of scientific knowledge. These writers imply that a better measure of the growth of knowledge would be to take the increase in the numbers of heavily cited publications per year. As will be shown in Chapter 4, however, citation is a social as well as an intellectual phenome-

11. Self-citations were excluded since the purpose of the analysis was to examine the processes affecting the transmission of ideas from one individual to another.

non. The extent to which publications are cited is related to the stage of development of a research area and cannot be used as a simple measure of its growth. Alternatively, if the sociologist attempts to measure growth by identifying important publications on the basis of historical studies, for example, he risks making highly subjective judgments.

The use of citation linkages between scientific papers is an approximate rather than an exact measure of intellectual debts. Little is known about how scientists decide to cite papers in their work and presumably not all of the citations in a particular paper have contributed equally to its contents. Some citations are to papers that played a central role in the development of the author's ideas; others are to papers that played only a peripheral role. Sociometric choices can also be criticized as unreliable indicators of relationships between scientists since it is obvious that the scientist may not recall all such contacts and may be biased toward reporting contacts with more prestigious individuals and ignoring those with less prestigious individuals.

In the absence of other equally good measures, however, I will proceed on the assumption that such approximate measures can be used to provide some indication of the actual nature of relationships among scientists in a research area. Once those relationships have been delineated on such a basis, guidelines may be provided for an analysis using more reliable indicators.

Using these indicators of social and intellectual relationships, I will show that social organizations develop that contribute to the logistic growth of knowledge in scientific research areas. Chapter 2 will demonstrate that social interaction plays a role in scientific growth, and will discuss models of intellectual change that are compatible with such a finding. On the basis of this discussion, a model of scientific growth will be presented that explains the logistic growth of research areas in terms of a series of changes in the characteristics of scientific knowledge in a research area and of the scientific community that is studying the research area. Figure 1 summarizes this model.[12]

12. See the appendix to this volume for figures, tables, and sample questionnaire.

Subsequent chapters will review the evidence that exists to support the model. Chapter 3 will present data concerning various types of social organization that emerge in research areas. It will show how certain types of social organization contribute to the rapid and efficient development of scientific innovations. Chapter 4 will examine the role of personal influence in the diffusion of scientific knowledge and the selection of problems for research. In Chapter 5, variations in the pattern of scientific growth and what such variations imply about the cognitive and social organization of such research areas will be discussed. Chapter 6 will treat the interrelationships, both cognitive and social, that link research areas into a complex pattern and permit the diffusion of ideas among them.

In Chapter 7, the implications of the model of scientific growth for improvements in the speed and efficiency of scientific communication will be discussed. Chapter 8 will explore the hypothesis that there are similarities between the communities that produce scientific ideas and those that produce, for example, artistic, literary, and religious ideas. Prospects for the convergence of the sociological analysis of these topics will be discussed.

2
Scientific Communities and the Growth of Knowledge

A FIRST STEP IN ANALYZING THE EFFECT OF SCIENTIFIC communities upon the growth of knowledge is to show that social interaction plays a role in scientific growth. If this is the case, what model of intellectual change is compatible with this finding and also fits the pattern of growth of publications and new authors that is exhibited by research areas? Three models of scientific growth will be described and evaluated according to these criteria. For the model that best fits the data, the nature of the cognitive and social events that occur during the growth process will be discussed.

Scientific Growth as a Diffusion Process
The fact that scientific literature in research areas exhibits a period of exponential growth (see chap. 1) indicates that scientific growth is a social process as well as a cognitive one for the following reason: If scientific growth represents the accretion of many small innovations, and if, in producing these innovations, authors are indeed building upon each other's work (as analyses of their citations to each other's publications suggest), then it would appear that such authors are adopting some of each other's innovations. In this sense, the growth of scientific knowledge is a kind of diffusion process in which ideas are transmitted from person to person. The diffusion of many types of innovations has been

shown to follow the logistic growth curve (Rogers 1962). In these studies, the exponential increase in numbers of adopters has been explained as a social influence process. When members of a social system are communicating with one another, a kind of "contagion" effect occurs in which individuals in a social system who have adopted an innovation influence those who have not yet adopted it.[1] The probability that a member of such a social system will adopt an innovation increases over time because it is related to the number of people who have already adopted the innovation (Coleman et al. 1966: ch. 7). As a result, the number of individuals adopting an innovation increases exponentially for a time. When individuals in a system are not in communication with one another, the probability that a member of the system will adopt an innovation remains constant and the pattern of growth is linear. Coleman and his associates have demonstrated the existence of both types of diffusion using two samples of doctors, one linked by sociometric ties and one not so linked.

Thus the exponential growth of scientific knowledge can be interpreted as a "contagion" process in which early adopters influence later adopters, which in turn creates an exponential increase in the numbers of publications and the numbers of new authors entering the area. The rate of expansion will vary depending upon the number of people with whom each scientist has personal contact. A small change in this number can greatly affect the total number of people receiving information after any specified number of exchanges has taken place. If the average scientist is in communication with three other scientists, after three exchanges of information, 22 scientists will have been in contact with each other

1. For the growth of a research area one could use the approach described by Stone (1966: 111) in his analysis of the demand for places in education. Stone suggests that "higher education should be regarded as a series of epidemic processes in which changes in the demand for places depend, in part, on the number already infected and so liable to infect others and, in part, on the number not yet infected and so available to catch the infection." Stone shows that the system of equations derived from his hypotheses gives rise to growth curves. For another approach to this problem see Goffman (1966).

(see Table 1). If the average scientist is in communication with six scientists, 187 scientists will have been in contact after three exchanges of information. Of course, this model assumes that all scientists have the same number of personal contacts, which is a useful approximation rather than a strictly accurate representation of reality. At any rate, the degree to which scientists in a field are in a position to exchange information through personal contact (i.e., correspondence, telephone, meetings, conferences) can affect the spread of information and the likelihood that a new idea will be widely adopted.

Periods of exponential growth can occur because research areas, although generally small, are capable of being expanded at relatively short notice if scientists with secondary and tertiary commitments to an area decide to shift their research priorities. In other words, if a particular field becomes especially attractive, a pool of scientists somewhat on the periphery can be rapidly assimilated. In addition, new scientists are continually being trained who are relatively free of previous commitments and thus available to exploit promising areas.

In order to test the hypothesis that science grows as a result of the diffusion of ideas that are transmitted in part by means of personal influence, the growth rate of research areas in which scientists were known to have been interacting with each other and with scientists who had not previously published in the area was compared with the growth rate of research areas in which they were known not to have been interacting in this manner. When members of a research area are interacting with other scientists, there should be a period of exponential growth since the probability that scientists who have not previously published in the area will adopt an idea will increase in proportion to the number of people who have already adopted it if these individuals are in communication with one another. This should be reflected in the growth of numbers of publications and numbers of new authors publishing for the first time in the research area. When members of a research area are not interacting with each other and with scientists who have not published in the area, the growth rate should be linear since the probability that one of the latter will adopt an innovation will remain constant at all points in time.[2]

In the two areas (the diffusion of agricultural innovations in rural sociology and the mathematics of finite groups) that I studied, it was clear that interaction was occurring among the members (see chap. 3). They were aware of each other's existence, and mutual influence and communication occurred. In both these areas, cumulative growth of publications and of authors entering the field followed the characteristic pattern of the logistic curve (see Figs. 2–5). In a third field, the phage area of molecular biology, which was also characterized by interaction and personal influence (Mullins 1968c), the cumulative numbers of new authors entering the field also fit this curve (see Fig. 6).

In contrast to these findings, in two fields in which the level of interpersonal communication and influence was low, the cumulative growth of publications was approximately linear (see Figs. 7 and 8). The first was an area of mathematics, the study of invariant theory. Fisher (1966, 1967) has documented the failure of members of the field to recruit students with a strong commitment to the field. This was in part due to the fact that many of these mathematicians happened to be located in academic institutions where graduate students were not being trained. Fisher argues that a topic is not perceived as important solely upon intellectual grounds. Without intellectual "promoters," it loses its ability to compete with other areas for adherents.

The second area in which personal influence was absent was an applied behavioral science, reading research (Barton and Wilder 1964). A study of a segment of this field showed that lack of interpersonal communication seriously weakened the intellectual development of the field. The same problems were repeatedly selected by successive generations of authors;

2. It has been suggested in another context (Orr and Leeds 1964) that exponential growth in science is more apparent than real on the grounds that new materials are continually being redefined as relevant to any particular area. In research areas in which the bibliographies were compiled after exponential growth had stopped (see fig. 2–5) this argument would not hold.

knowledge failed to cumulate. These authors showed that the absence of communication networks was due to the fact that most members of the field had a very limited contact with it. They wrote theses in the area and afterward did no further research in the area. As a result, the field lacked individuals who had a long-term commitment to it and who could recruit and train new members and influence their selection of relevant research problems. The fact that the same problems were repeatedly selected by successive generations of authors suggests that ideas diffuse more effectively when transmitted by individuals rather than by publications alone.

These findings show that scientific growth is both a social and a cognitive process. Social interaction facilitates the diffusion of ideas that in turn makes possible cumulative growth of knowledge in a research area. In the following sections, I will discuss the relationship between models of cognitive growth and patterns of social interaction.

Models of the Growth of Knowledge

The fact that the growth of research areas fits the logistic curve suggests that there are changes in the social relationships among scientists during the process of growth. This raises the question of the relationship between these changes and the process of intellectual development in research areas. Are they interdependent? The various phases of scientific growth have been conceptualized in several ways.

Probably the most widely accepted model of the growth of knowledge is one that views it as a cumulative progression of new ideas developing from antecedent ideas in a logical sequence. Hypotheses derived from theory are tested against empirical evidence and either accepted or rejected. There is no ambiguity in the evidence and consequently no disagreement among scientists about the extent to which an hypothesis has been verified. Many discussions of the nature of scientific method are based on this model of scientific growth.

An alternative model that has been applied most frequently to the growth of nonscientific knowledge states that the origins of new ideas come not from the most recent developments but from any previous development whatever in the history of the field. In this model, there is a kind of random selection across the entire history of a cultural area. If this were the case in a research area, a scientific innovator could take his inspiration from work that had been completed in any previous century, and contemporaries would not necessarily build upon each other's work. Price (1970) argues that this kind of highly unstructured growth is characteristic of the humanities.

The first of these models stresses continuous cumulative growth, the second its absence. Another type of model includes periods of continuous cumulative growth interspersed with periods of discontinuity. One version of the latter is Kuhn's analysis of scientific change (1962) that combines periods of cumulative growth, which he calls normal science, with periods of crisis or revolution. In his view, "normal" scientific activity in a research area is guided by a paradigm that defines the fundamental problems. The attention of scientists is directed toward these problems exclusively. As a result, scientific knowledge grows in a systematic fashion, building upon previous work. The task of the scientist is that of puzzle-solving rather than that of searching for entirely new scientific innovations in his research area.

In time, however, phenomena that the paradigm cannot explain become increasingly important. When these anomalies can no longer be ignored, the field goes through a period of crisis while the old paradigm is under attack and a new one is sought. A new paradigm is generally resisted, particularly by older scientists, until it has proven its superiority over the previous one. When the old paradigm has finally been rejected, much of the work that it stimulated is no longer considered relevant and ceases to be part of accepted knowledge in the field.

Another version of this model describes a period of growth ending in exhaustion of the ideas that had stimulated the growth. For example, in Kroeber's view (1957), concentration of interest among scientists upon a particular set of problems leads to a kind of crescendo of activity in which the importance of the results obtained steadily increases and then gradually tapers off as the potentialities in the methods being used to solve the problems are exhausted or as the problems are solved. The process is repeated when a new orientation toward these problems is discovered or when an entirely new set of problems is defined. The information produced by the previous approach is reinterpreted rather than being completely discarded.

Toulmin (1963, 1967) on the other hand sees a process of continual change or "micro-revolution" that results in the "selective perpetuation of preferred intellectual variants." This is a frankly Darwinian model. Ideas that are applicable to more than one field have the greatest likelihood of survival. Toulmin shares Kuhn's view that there are periods in science when knowledge does not cumulate. He calls them "recurrent periods of self-doubt," during which scientists tend to question whether science can explain anything. These crises are resolved by changes in the underlying assumptions of the field. In contrast to Kuhn's view that an old model is discarded, Toulmin argues that an old theory may survive in a new field or be reintroduced into a field at a later date as new facts are discovered.

To some extent, Holton's conception of scientific change (1962) incorporates both Kroeber's and Toulmin's interpretations. Science as a whole grows because individual fields grow. Each one is eventually exhausted, but new ones are continually emerging due to the discovery of linkages or connections among old ones. Like Toulmin, he sees the birth and death of scientific fields in terms of the biological evolutionary analogy, of a struggle for survival among ideas. Certain central concepts, utilized in a wide variety of fields, provide continuity, but the multiplicity of effort coupled with the essential democracy of scientific institutions that permits every idea a hearing ensure endless opportunities for the production of new ideas that then compete for survival.

Unlike the first two models described, this type of growth model includes the notion of a special kind of cognitive event that plays an important role in bringing about scientific change. Kuhn relies most heavily upon this factor. In his model, a "paradigm" makes possible a period of cumulative growth in a scientific field. His definition of this phenomenon has aroused considerable controversy among historians and philosophers of science, largely because of its ambiguity. Masterman (1970), who performed a content analysis of Kuhn's discussion of the concept (Kuhn 1962), shows that his definitions of a paradigm can be separated into three categories: (1) metaphysical paradigms, in which the crucial cognitive event is a new way of seeing, a myth, a metaphysical speculation; (2) sociological paradigms, in which the event is a universally recognized scientific achievement; and (3) artifact or construct paradigms, in which the paradigm supplies a set of tools or instrumentation, a means for conducting research on a particular problem, a problem-solving device.

Masterman argues that the last of these three is the meaning that is most suitable to Kuhn's view of scientific development.[3] In other words, scientific knowledge grows as a result of the invention of a puzzle-solving device that can be applied to a set of problems producing what Kuhn has described as "normal science." According to this definition, a paradigm is not identical with either a theory or with a technique. The puzzle-solving device precedes the development of theory, and it is more than simply a method of doing research. A method can be applied to many sets of problems. The true puzzle-solving device appears to be highly specific in its application to a particular set of problems. According to Masterman, when this device is applied to problems beyond its range, anomalies occur and its limitations become evident. This sets in motion a chain of events that leads to the development of a new paradigm.

The crucial cognitive events that are described by other writers, however, are much more similar to what Masterman calls a metaphysical paradigm than to the artifact paradigm or puzzle-solving device. Toulmin, for example, speaks of an expectation regarding scientific phenomena, an intellectual pattern that defines the range of things that can be expected. Like Kuhn, he stresses the effect of such an orientation upon those who accept it: "We see the world through [the funda-

3. Kuhn in a recent review of the concept (1970) analyzes it in a similar fashion with the same emphasis.

mental concepts of science] to such an extent that we forget what it would look like without them: our very commitment to them tends to blind us to other possibilities" (1963: 101).

Holton (1964) speaks of central concepts that are shared by many scientific fields and also of "themata" that influence the type of data that the scientist gathers and the way he analyzes it. Themata are fundamental presuppositions that are neither objectively observed nor quantified but are remarkably persistent in scientific thought. He argues that some of the themata of Greek science continue to influence modern science, such as a belief in a principle of potency, the ancestor of the modern concept of force.

Back (1962), building upon Kroeber's model, suggests that development of a research area begins with a synthesis or integration of previous ideas that results in a "pattern" that is then explored in considerable depth. Finally, Mulkay (1970), in a critique of Kuhn, concludes that the term "paradigm" refers to particular scientific achievements but that, in order to understand how a paradigm operates, supplementary concepts are needed, specifically cognitive and technical norms that define "legitimate problems and acceptable solutions" on the one hand and "preferred types of instrumentation" on the other. Mulkay appears to have in mind something that is closer to the artifact paradigm or puzzle-solving device than to the metaphysical paradigm.

This discussion suggests that there are in fact two facets to the cognitive event that stimulate the growth of knowledge: the paradigm as a way of seeing, a perspective, a pattern, and the paradigm as a special kind of tool or problem-solving device. Intuitively, it seems plausible to argue that the paradigm as a way of seeing or interpreting events is shared by large numbers of research areas. Entire disciplines sometimes share the same orientation toward phenomena. On the other hand, what Kuhn and Masterman have in mind regarding a puzzle-solving device appears to be something that is highly specific to a research area. The "strength" of the puzzle-solving paradigm comes from its unique applicability to a particular set of problems. It is precisely this characteristic that makes the paradigm so difficult to define in general terms.

Evaluation of Growth Models

In the previous section, three models of scientific change were described. Which one is the most appropriate as a guide to the sociological analysis of scientific growth? The first model has been accepted for decades as an accurate description of the natural sciences. It has recently been questioned by a number of writers because it does not explain the existence of certain types of behavior that are observed among scientists. For example, this model assumes that scientists agree about the interpretation of scientific data and thus scientific controversy is unnecessary. It does not include the possibility that two scientific theories that are mutually incompatible could be used to interpret the same data. This model also assumes that scientists at all times possess all the information they need to evaluate a new finding or a new theory. Opponents of the model argue that scientists are often forced to commit themselves to ideas before all the evidence is or can possibly be available to support these ideas. It has recently been argued that one of the factors that makes possible the cumulative development of knowledge in the natural sciences is the use of theoretical constructs that are not completely defined when they are first used: "When the construct is first formulated, the theorist simply does not yet know all the assertions which he may wish to make" (Willer and Webster 1970: 753). A complete definition of such constructs is the product of a long period of subsequent empirical and theoretical work during which new information is assimilated into the definitions of the constructs and into the theories themselves.

Dolby (1971) has pointed out that the nature of the evidence that scientists actually use to support their hypotheses is rarely as precise and rigorous as this model assumes. Scientific assertions are not accepted on the basis of logical reasoning alone but also on the basis of how well the paper fits into a complex set of explicit and implicit assumptions that are not empirically testable. Fields in which methods are so precise and data so unambiguous that hypotheses are accepted after a single test, the so-called crucial experiment, are the exception rather than the rule. An example of one such hypothesis is that of the nonconservation of parity in physics. The failure of parity to hold in all cases was accepted by the relevant scientific community immediately after its demonstration. Yet earlier proof of this phenomenon, obtained before the development of a theoretical explanation for it, was largely ignored (Hagstrom 1965: 253). It is much more common for the scientific community to withhold acceptance of an hypothesis until it has been subjected to a number of tests or replications, often extending over a considerable period of time. As Kuhn (1962) has stressed, some findings may be ignored by the scientific community because they do not fit the current paradigm (for a series of examples of this phenomenon in the career of a recent Nobel prizewinner see Dessler 1970). If the phenomenon is very complex, findings may be contradictory and difficult to evaluate for considerable periods of time (for example, see the discussion of the history of molecular biology by Hess 1970). During this period, scientists frequently disagree among themselves regarding the extent to which an hypothesis has been verified.

The new model of scientific change, several variants of which have been described above, allows a much greater degree of ambiguity, controversy, and discontinuity in the development of scientific knowledge. Science is not a continuous line of development but is marked by numerous shifts from one theory to another and breaks in continuity in which subjects are neglected for a period and then reinvestigated from a new point of view. In a sense, the first model can be viewed as a special case of the new model. It describes what happens during one of the phases of growth. Normal science as described by Kuhn can be explained in terms of the cumulative model.

Another argument against the first model is that it does not fit the pattern of growth of the scientific literature and of scientific authorship. Examination of the cumulative numbers of new publications and of authors publishing for the first time in research areas shows that growth of research areas fits the logistic curve. The first model does not imply that research areas go through phases of growth of this sort. Instead it implies continuous growth at the same rate at all times. The second model, which describes new developments emerging from any previous development, also does not fit this pattern of growth since it implies irregular growth at all times. The third model, which describes periods of growth followed by periods of discontinuity or decline, fits the growth data very well.

The second model is also contradicted by data concerning the connections between pieces of scientific literature as evidenced by their citations. Studies of citations (Price 1965a) reveal that references in the scientific literature are about equally divided between those to a group of relatively recent papers and those to a random selection of the remaining literature, suggesting a concentration of interest upon immediate predecessors in a research area. If scientific growth resembled the "random" model, citation patterns would be unlike those actually observed.

Finally, in evaluating these models, their implications for patterns of social interaction in research areas must be considered and compared with what we know about scientific communities. The second or "random" model implies very little social organization in research areas. If, at any point in time, scientists are drawing upon the entire archive of literature, one would not expect a sufficient concentration of their interests to allow the mutual exchange of ideas or even constructive controversy.

While the cumulative model of scientific growth is not incompatible with the presence of social interaction among scientists, it does not imply a high level of solidarity among them since conflict and controversy are at a minimum. It is during periods of controversy that solidarity among opposing groups of researchers is likely to appear or, alternatively, during a period when a new field is being developed. For example, one would expect considerable solidarity to develop between researchers at points when a break is being made with past work and a new theoretical approach is being created. The tension involved in rejecting an old point of view and replacing it with a new and somewhat speculative one would seem likely to produce strong ties between those who support the new viewpoint in opposition to those who continue to support the older viewpoint.

Interaction between Cognitive Growth and Social Organization

It has been shown that exponential growth of a research area reflects a social interaction process in which contact between scientists contributes to the cumulative growth of knowledge. This suggests that there is an interaction between cognitive events and social events in the development of a research area that parallels the stages of the logistic curve. In the early stages of growth of a research area, an interesting cognitive event (possibly along the lines of a paradigm as described by Kuhn) occurs and attracts new scientists to the area that previously has been sparsely populated. This is followed by a period of cumulative development of knowledge in the area (in Kuhn's terms, normal science) during which the number of publications and of new authors entering the field grows exponentially.

The shift to exponential growth is marked by the appearance of new groups of scientists within the research area. These groups can be described from two points of view: (a) their structural aspects: who is linked to whom by sociometric ties; (b) their normative aspects: the types of attitudes and behavior that are expected of their members.

Recent studies of research areas (see chap. 3) show that there are two distinct types of subgroups. One type consists of groups of collaborators. For example, Price and Beaver (1966), in their analysis of memos contributed to an information exchange group that had been formed to speed the rate of communication of research findings in the research area concerned with the study of oxidative phosphorylation, found that on the basis of published collaborations the entire set of authors formed numerous nonintersecting subgroups of various sizes. The larger ones contained a few very productive scientists and many relatively unproductive ones. Other studies show that these groups are linked to one another through their leaders who communicate with each other and transmit information informally across the whole field. This enables them to monitor the rapidly changing research "front" and to keep up with new findings during a period of rapid growth. Thus the second type of subgroup in a research area is a communication network or "invisible college" that links groups of collaborators.

It appears that the large groups of collaborators and the communication networks that link them play very different roles in the development of the area. Under the leadership of one or two scientists, the groups of collaborators recruit and socialize new members and maintain a sense of commitment to the area among existing members. Mullins (1968b) has called these groups "solidarity groups" in order to emphasize their role in promoting dedication to an area. The leaders of these groups also define the important problems for research in their areas. As we shall see in Chapter 4, their interpretations of seminal works influence the subsequent research of other members in the field.

Since most scientists produce only one or two publications in their careers, it is necessary to recruit many individuals to an area in order to locate the few who will become highly productive. Some scientists are strongly motivated to surround themselves with collaborators and students in order to produce a greater volume of work and to provide for continuation of their work after they have retired. For example, one particularly strong personality in the phage area of molecular biology attracted students to the area, set guidelines for their research, and obtained resources for its continuance (Cairns et al. 1966; Mullins 1968b and c). The atmosphere of the group is indicated by the following quotation from an autobiographical account by one of the members:

Once started with the bacteriophages, I continued research on them because it seemed that here one could do more interesting experiments than with any other material. Furthermore, the engaging group of people working on these viruses was highly compatible; among ourselves we saw, and still see, very little competitive secrecy and backbiting; we were all interested in cooperating with each other to promote the work as a whole (Cairns et al. 1966: 77).

Griffith and Miller's studies (Project on Scientific Information Exchange in Psychology 1969) of the social organization of research areas in psychology also point to the importance of some individuals in the development of an area. For example, in one area, which was characterized by a high level of informal communication, they found that a senior researcher-teacher had trained a considerable proportion of the active researchers in the field, raised and managed research fellowship funds to support students, and was regarded as having set standards for both research and theory in the field. When this man left the area, a younger man assumed a similar role.

Many of the characteristics of a solidarity group are exemplified by the Bourbaki, a group of French mathematicians that was active for over thirty years (Halmos 1957): (1) it developed radically new approaches to its field; (2) members were highly critical of each other's work; (3) morale was high—there was a great deal of *esprit de corps,* a high value placed upon humor and practical jokes of an intellectual nature, rapid conversation, and communal enjoyment of fine French cuisine; (4) there was a clearly defined "style of work," which was characterized by periodic meetings to review manuscripts; (5) the group had several leaders who had considerable prestige in their discipline and who maintained their commitment to the group for long periods of time; (6) there was constant turnover among the less prestigious members of the group; (7) in its heyday, it had no difficulty in recruiting new members; and (8) its activities tended to be centered at one or two institutions where the leading members resided, although many members were dispersed.[4]

4. An unusual feature of this group was that it published under a pseudonym, Nicolas Bourbaki (Halmos, 1957). Its prestige was so high that merely being identified informally with the group was such an honor for a mathematician that he was able to forego the usual formality of having his name attached to his publications. The names of the members were known to most mathematicians who were doing related work.

As further development of a research area becomes increasingly difficult, one would expect the area to become less attractive to those outside it so that exponential growth in the expansion of the field is replaced by a period of linear growth. Kuhn has described such difficulties as anomalies that the paradigm is unable to explain. One would anticipate that this period of cognitive uncertainty would lead to increasing defensiveness on the part of different subgroups concerning their own interpretations of the intellectual problems of the area. This would be likely to produce increasing differentiation between groups of collaborators and a decline in the exchange of ideas among them. The communication network or invisible college would function less effectively than before.

For example, an informant in a research area in physiology suggested that conflict between groups of collaborators was a characteristic of an area that had been active for a considerable period of time. In the early stages of rapid growth, the productive scientists have not had time to develop sizable groups of colleagues. Interaction between them and other members of the area is unrestrained. After a number of years, these same scientists have established themselves at the centers of clusters of collaborators and students. They tend both to defend their own ideas and to resist ideas put forward by newcomers. Sometimes they discount new ideas as being not really new. This informant suggested that "one tends to think that the relevant issues have been solved if one has been in a field for a long time." Such tendencies make it more difficult for an area to recruit new members, thus contributing to its decline.

A tendency to resist new developments can affect even the most eminent scientists in a research area. Gamow (1966: 32–33) in his history of quantum theory tells how the revolutionary ideas of the young Niels Bohr were resisted by the eminent J. J. Thomson whose ideas had been important to a previous generation. Bohr found it impossible to develop his ideas in Thomson's laboratory and moved to another and more hospitable scientific environment.[5]

5. For other examples see Barber (1961).

During a period of crisis when a new approach is being formulated in opposition to a previous approach, one would anticipate a decline in growth. If a new approach is accepted, a new cycle of growth follows. Krantz (1965) found a decline in numbers of publications appearing in two areas of psychology following discoveries that could not be reconciled with existing theories. Neither field produced a new approach that could reverse its decline.

This discussion has been phrased in terms of Kuhn's model in which episodes of crisis close periods of rapid growth and are followed by the acceptance of new paradigms and new periods of growth. Variations of this model emphasize the exhaustion of paradigms, at least in terms of their applicability to particular topics. Scientists abandon such fields in favor of new ones that are opening up.

According to Kroeber's version of this model, some form of synthesis or integration of knowledge produces a period of rapid growth in a research area, during which facts are accumulated to test the theory. One would expect that social organization would emerge in the area as it expanded in size. As a result of their exclusive concentration upon these problems, the interests of members of the area would become increasingly specialized. Once the original problems have been solved, members of the area are unable to define new ones because of their restricted range of interest. The effects of exhaustion of the model upon the social organization of the area are more difficult to predict. If, as Back (1962) suggests, the interests of members of the field become increasingly narrow as they test and refine their ideas, it seems likely that subgroups in the area would become increasingly specialized in narrow segments of the field and thus increasingly autonomous.

There has been some discussion about which of these versions of scientific change, revolution or dissolution, most accurately describes modern science. Kuhn himself has noted a change in the character of scientific revolutions in physics. Since the middle of the nineteenth century, the high degree of quantification in that discipline has meant that controversies can be resolved relatively quickly with a high degree of consensus.

Other writers (Ben-David 1964; Mulkay 1969) have argued that the proliferation of scientific fields has blunted the revolutionary character of science. An idea that is rejected in one specialty may be accepted in another. It is more difficult to correlate a particular specialty with a particular paradigm, so that rejection of a paradigm has less noticeable effects. Much scientific change is simply the discovery of entirely new areas to be studied and the application to them of paradigms that have proved their usefulness in other fields.

If this is the case, the evolutionary models of Holton and Toulmin provide a better explanation of contemporary scientific change than Kuhn's "crisis" model. The existence of hundreds of fields, growing and declining, linked to some extent by concepts that have proved useful in several areas and with no clear-cut boundaries between them, permits both rapid diffusion of ideas and also the coexistence of mutually incompatible ideas if applied to different research topics. These writers largely ignore the role of scientific communities in this process, but it is clear that they must play important roles in the transmission of information and in the definition of relevant topics for their members.

This same process can be seen on a much larger scale in the emergence of new scientific disciplines (Ben-David and Collins 1966; Oberschall 1968), an event that occurs much less frequently. On this level, several additional problems have to be solved before new knowledge begins to emerge. For example, the redefinition of research goals that is involved in the emergence of a new discipline is much more drastic than that which occurs in the selection of a new research area. Second, these scientists must usually develop new organizational positions in which to perform their new roles, and resources must be found to support the new discipline. On the level of the research area, these problems can be assumed to have been solved. Scientists move into new research areas within established disciplines and their organizational support is assured. In the case of new interdisciplinary research areas, this may not always be the case and such difficulties may inhibit their development, as will be discussed in Chapter 6.

Once the new discipline has been admitted to the appropriate organizations, its development follows a course not unlike that of a research area. Chains of teachers training students who in turn train new students are essential to the process. When students are not available, growth of the discipline is arrested.

Conclusion

Existing models of the growth of scientific knowledge describe the cognitive aspects of growth and generally ignore the role of scientific communities in this process. In this chapter, it was shown that exponential growth of research areas in basic science is indicative of a social process underlying scientific growth. The various stages of logistic growth of a research area are accompanied by a series of changes in the characteristics of scientific knowledge and of the scientific community that is studying the area. Interesting discoveries that provide models for future work (paradigms) attract new scientists to the area during stage one. In stage two, a few highly productive scientists set priorities for research, recruit and train students who become their collaborators, and maintain informal contact with other members of the area. Their activities produce a period of exponential growth in publications and in new members in the area. As the implications of the seminal ideas are exhausted or become increasingly difficult to test due to the appearance of anomalies that cannot be explained by the original model (stages three and four), new scientists are less likely to enter the area and old members are more likely to drop out, thus leading to a gradual decline in the number of new publications and in over-all membership in the area. Those who remain are likely to develop increasingly narrow and specialized interests as the possibilities for research dwindle or to be divided into factions on the basis of theoretical controversies. In the latter case, acceptance of a new paradigm leads to a new cycle of growth. Figure 1 summarizes these relationships.

3
The Social Organization
of Research Areas

SINCE SCIENTISTS IN RESEARCH AREAS CAN HAVE A NUMBER
of different types of social relationships with one another, it
is necessary to use a variety of indicators to measure social
organization. For example, many scientists discuss their ongoing research with other scientists in order to obtain advice
and also information about similar studies. Many scientists
collaborate with other scientists and publish the results
jointly. This represents a very intensive type of communication. Frequently, collaborators are teachers and students.
Even without formal collaboration, the teacher who trains
a student often retains a close relationship with him in later
years. In any case, the teacher's ideas and orientation toward
the field are likely to leave their mark upon the student's perception of the field. Finally, scientists are influenced by other
scientists and their publications in their selection of problems and techniques for their research. Often the men who
exert such an influence are teachers, although obviously in
other cases there may be no face-to-face contact between
the author of a seminal publication and those who are influenced by it. Thus, the most important indicators of social
organization in a research area are informal discussions of
research, published collaborations, relationships with teachers, and the influence of colleagues upon the selection of
research problems and techniques.

From each of these types of ties among scientists one obtains a somewhat different picture of the extent to which members of a research area are linked to one another. Nevertheless, if one uses the various indicators of linkage separately and then in combination, one is provided with a fairly complete picture of the amount of relatedness that exists. Some members may be related to other members through influences upon the selection of problems or techniques, others through some type of collaboration or through informal communication. Thus, if social organization does exist in a research area, most members should be related to others in at least one of these ways. Indirect ties are as important as direct ties, for a scientist who communicates with another scientist in this way may obtain information that was transmitted to the second scientist by a third.

Allen (1969b) has argued that it is unlikely that indirect ties between scientists reflect the actual flow of information and influence and that therefore it is meaningless to consider them. He points out that the number of intermediaries separating two individuals is inversely related to the probability of information being exchanged by them in this manner. It is important to realize, however, that a research area is not entirely a face-to-face group. Its members are widely dispersed geographically. Many exchange ideas by correspondence or through publications. Third parties play important roles in transmitting information between persons who may never actually meet. In the present study, a technique for locating indirect ties using matrix multiplication known as the Sociometric Connectedness Program was used (Coleman 1964: 447–55). In both areas, the number of intermediaries was small on most measures.[1] In addition, examination of citations appearing in the literature of one of the areas made it possible to validate the existence of influence transmitted through intermediaries.

1. In both areas, the majority of those having indirect ties were linked by not more than two intermediaries on most measures. When all the ties were examined together, a majority of the members of both areas had some indirect ties based on between five and nine intermediaries. But over two-thirds were linked to one of the most productive scientists (the sociometric stars) by not more than one intermediary.

The Research Area as a Social Circle

A social circle is characterized by the presence of direct and indirect ties among many but not necessarily all of its members. The extent to which members of an area are linked to one another can be measured: (1) by examining the proportion of relationships that actually occurred in relation to all of those that possibly could have occurred. In other words, if every scientist was linked to every other scientist in the group, 100 percent of the possible relationships would exist; (2) by compiling sociometric networks showing how many of the scientists in the group are linked to one another directly or through a common link with other individuals.

PROPORTION OF POSSIBLE RELATIONSHIPS ACTUALLY OCCURRING

Dividing the number of direct and indirect ties that actually occurred by the number of ties that possibly could have occurred among members of an area provided an indication of the extent to which the members of the area were connected with one another.[2] The figures in the last columns of Tables 2 and 3 (see Appendix) show that the level of connectivity varied considerably depending upon the type of relationship examined.[3] A substantial proportion of connectivity appeared only when all types of ties were considered simultaneously (i.e., when the ties between individuals could be any of the several that were studied). This suggests that social organization in a research area is revealed only when a variety of different types of relationships between members of a research area are examined.

Connectivity as measured here is a very stringent measure of total relatedness in an area. This type of connectivity measures relatedness on the basis of choices in one direction at a time. This means that a scientist's links to the network

2. The number of possible ties is computed by multiplying the number of members in the area by the same number minus one, thus eliminating self-choices.

3. Only respondents are included in these tables. One type of tie, influence on the selection of research techniques, was not assessed for the mathematicians since it appeared to be inappropriate in terms of their type of research activity.

are measured either on the basis of his own choices of others or others' choices of himself but not both simultaneously. He then receives a score based upon the number of such direct and indirect links. If he chooses an individual who does not in turn choose anyone, his score is low even though the other individual may be linked to the larger network by someone else's choice. Thus, in groups of scientists where a large proportion of choices are directed toward a few leading figures who reciprocate only a few of them, this type of connectivity is low.

SOCIOMETRIC NETWORKS

A less stringent measure that has been used by Crawford (1970a) is the size of the network linking those who had published in an area on the basis either of their own choices or those of others who had published in that area. For example, information that respondents gave concerning the identities of the persons with whom they discussed their ongoing research informally was used to construct a sociogram showing all their sociometric choices and, on this basis, 66 percent of those who were continuing to do research in the mathematics area were members of a single network during the period when questionnaires were mailed to members of this group. In the rural sociology area, 73 percent of those who were working in the area at the time that group was studied were linked to such a network.[4]

Respondents were also asked to name persons or publications that had influenced their selection of research problems in these areas. When this type of influence was used as the basis for choices, 70 percent of the members of the mathematics area were linked in one large network, and 43 percent of the members of the rural sociology area were so linked.[5]

---

4. Only individuals who reported on the questionnaires that they were continuing to conduct research in these areas were included in these two sociograms.

5. These figures and those in the next paragraph include nonrespondents for both areas ($N = 102$ for mathematics and 221 for rural sociology). The comparable figures excluding nonrespondents are: (a) selection of research problems network: mathematics—72

When several different types of ties were considered simultaneously, these figures were higher. In the mathematics area, considering informal discussion of research, influence on the selection of research problems, and ties based on student-teacher relationships, published collaborations and collaboration on work in progress, 78 percent of all the members of the area were linked in a single network. In the rural sociology area, 74 percent of the members were linked on the basis of these various types of ties.[6]

Using this criterion of relatedness, there appeared to be in each of these areas a sizable core of individuals who were connected with one another. This is a reasonable outcome in science where students or collaborators of very productive scientists are brought into contact, directly or indirectly, with many other scientists in the field. The accuracy of sociometric data of this kind can be questioned, of course. Respondents may not recall their contacts accurately. Alternatively, they may tend to include persons of higher status than themselves and omit persons who are of the same or lower status. In addition, we are assuming that persons linked by third parties are capable of influencing each other. At least partial validation of these results was obtained through an analysis of citations appearing in the literature of the mathematics area. The network linking members of that area on the basis of the references to each other's work that appeared in their publications was very similar to that produced by the direct and indirect links generated from the names mentioned by members of the area regarding influences upon their selection of research problems. In each case, one large network of individuals linked directly or indirectly emerged. Eighty-three percent of the members of the area were linked to such a network on the basis of their own references or the citations of others. Seventy percent were linked to such a network on the basis of sociometric

---

percent; rural sociology—62 percent; (b) total ties network: mathematics—81 percent; rural sociology—86 percent ($N = 64$ for mathematics and 147 for rural sociology).

6. One other type of choice was used here: influence on the selection of research techniques.

choices: naming or being named as influences on the selection of research problems in the area. These data suggest
that asking scientists to name influences upon their work
produces information that is fairly similar to that obtained
by examining influences using citations. When membership
in the two networks constructed by these different methods
was compared, it was found that the two methods agreed in
locating members either inside or outside the main network
in 71 percent of the cases.[7]

The significance of informal communication in a research
area is suggested by the fact that, among those who were
currently conducting research in these areas, 44 percent of
the rural sociologists and 51 percent of the mathematicians
rated their informal communication contacts as being very

7. Connectivity was substantially higher when based on members'
citations of each other than when based on their recollection of
influences upon their work. On the basis of others' citations of
group members, 29 percent of the possible ties in the area were
actualized. Sixty-two percent of the possible ties linking the High
Producers to the area were actualized compared to 46 percent of
those linking the Moderate Producers and 22 percent of those linking the Low Producers. The corresponding data for influences upon
the selection of problems was 4 percent, 19 percent, 7 percent, and
2 percent with nonrespondents included. The differences between the
two distributions are at least partly the result of having information
about all members of the sample for citation choices. Information
about influences upon the selection of research problems was available from 48 percent of the members of the mathematics area.

The question requesting names of scientists who had influenced
the respondents' selection of research problems was not answered by
41 percent of the members of the rural sociology area. A similar
question requesting names of scientists who had influenced their
selection of research techniques was not answered by 59 percent. In
general, failure to respond to these questions was related to low
productivity and lack of commitment to the area. A few scientists
indicated that the research topic had been assigned to them. Other
respondents claimed that these questions were difficult. The degree
of perceived difficulty seemed to depend on whether the respondent
thought that he was being asked to list every possible influence or
only those that had been especially important (the latter was what
was requested). In the lengthy career of a scientist, influences on the
development of research can be numerous, but it is quite possible
that only a relatively small number of names are very significant in
any particular research area. Some respondents seemed to have
difficulty or were unwilling to make the effort to distinguish such
influences.

important for their research. Several mathematicians who were interviewed by telephone made comments similar to this one: "You do your research alone, but if you were never able to talk to other people, you'd be in bad shape. You have to know what's up in the field for your research to have an impact."

For some members, however, contact with other members of these research areas had been very brief. Some of the less-productive members appeared to have been drawn almost involuntarily into the research activities of these areas and to have detached themselves as quickly as possible. Comments on the questionnaires implied that some of these individuals were unaware of any tradition of research in the area. (The problems associated with doing research in an area when the researcher is relatively isolated from other members of the group will be discussed later in the chapter.)

Results of Other Studies

Other studies of research areas have used different types of samples and have analyzed their data with different techniques. Nevertheless, findings tend to corroborate those reported above. Three studies, those by Crawford (1970a), Gaston (1969), and Lingwood (1968), have attempted to obtain sociometric information from a high proportion of researchers in particular research areas.

Crawford asked 218 scientists to name those with whom they communicated frequently and informally concerning their research on sleep and dreams. One large network dominated the area: 72 percent of the scientists were linked to one another by either direct or indirect communication ties. Crawford also found a high correlation between her sociometric data on informal communication and data obtained from an analysis of citations appearing in the literature of the area.

Gaston, who studied high-energy physicists in Britain, found that a subgroup consisting of about 30 percent of the entire group were in communication with one another. All of these individuals were linked to one another directly or indirectly.

Lingwood examined sociometric ties among a group of scientists in educational research. Preliminary findings indicated that researchers tended to mention other scientists within the same research areas in that specialty, although there were some areas in which this phenomenon did not occur. Griffith and Miller (1970 and Project on Scientific Information Exchange in Psychology 1969), who examined communication ties among the more productive scientists in several research areas in psychology, found that in some areas members chose each other frequently while in other areas they did not.

The failure to find evidence of social organization in research areas can be interpreted in different ways. It may mean that the categories that were chosen to designate a research area are not those currently being used by scientists to label their work. Lingwood suspected that his negative findings were due to faulty definition of research areas. Alternatively, it may mean that the area under consideration has not yet experienced the rapid growth that is associated with the development of social organization, or that it has experienced such growth in the past and is now a declining field or one that has been abandoned by all but a few diehards. For example, the areas studied by Crawford and by Gaston had recently had or were experiencing periods of rapid exponential growth. The absence of social organization in some of the areas that Griffith and Miller studied may be due to the fact that these areas were no longer "hot" areas. Although a number of publications may exist in an area, the interest of researchers may have dwindled so that the literature represents an empty shell from which active communication among researchers has vanished.

In other instances, the absence of social organization in a research area may mean that the research role has not been institutionalized in the discipline to which it belongs. When social contacts among researchers in a discipline are lacking, social organization is unlikely to emerge in research areas within the discipline (Barton and Wilder 1964).

The Role of Invisible Colleges

Analysis of the networks showed that anyone choosing even one of the most productive members of the research areas studied by the author could have been in contact with a large network of individuals. In other words, the high proportion of choices directed toward these individuals meant that members of these groups were not so much linked to each other directly but were linked to each other indirectly through these highly influential members.

Members of the rural sociology area were divided into subgroups on the basis of the number of papers they had written in the area and their present commitment to it. The latter was measured in terms of whether or not the individual was conducting research in the area at the time the area was studied by the author. Members of three subgroups indicated in response to an item on the questionnaire that they were continuing to do research in the area: (1) eight High Producers, each of whom had published more than ten papers in the area; (2) eleven Moderate Producers, who had published four to ten papers in the area; (3) thirty-three Aspirants, who had published less than four papers in the area. Members of two subgroups indicated that they had not continued to do research in the area: (1) nine Defectors, each of whom had published four to ten papers in the area; (2) eighty-six Transients, each of whom had published less than four papers in the area. No respondents who had published more than ten papers had ceased to do research in the area.

The much smaller size of the mathematics area did not make it practical to examine so many subgroupings. Instead, the area was divided into three parts, on the basis of productivity, using the same levels of productivity as indicated above. Four members of the area were highly productive, thirteen were moderately productive, and forty-seven were low in productivity. There was a strong correlation between productivity and present commitment to the area.

In the rural sociology area, 6 percent of the scientists received 58 percent of the choices.[8] Forty-six percent were not chosen at all. A similar distribution appeared in the mathematics area: 6 percent of the mathematicians received 38 percent of the choices while 34 percent were never chosen.[9] Those who were chosen frequently were the most productive scientists in these areas (see Tables 4–7). Examination of the proportion of relationships that actually occurred in relation to all of those that possibly could have occurred showed that the most productive members of both areas had more relationships with other members of their areas as measured by the direct and indirect choices of other members than did the less productive members of these areas (see Tables 2 and 3). Thus the connectivity that developed in this area was mainly the result of a large number of choices directed toward a few members.[10]

8. Nonrespondents are included. The following types of choices are examined: informal discussion of research, influence upon the selection of problems and techniques, relationships with teachers, and collaborative work in progress. The findings were similar when indicators were examined singly. For example, 6 percent of the members of the rural sociology area received 64 percent of the direct choices with respect to informal discussion of research and 69 percent of the direct choices with respect to influences on the selection of problems. Two percent of the members received 25 percent of the choices as thesis directors.

In a sample of high-energy physicists who were asked to name the two persons with whom they most frequently exchanged information, the proportion of physicists not named was slightly more than one-third (Libbey and Zaltman 1967: 31–32). In Gaston's sample of British high energy physicists, the proportion of physicists not named was 70 percent (Gaston 1969: 319).

9. Nonrespondents are included. The following types of choices are examined: informal discussion of research, influence upon the selection of research problems, relationships with teachers, and colloboration on work in progress. The findings were similar when indicators were examined singly. For example, 6 percent of the mathematicians received 37 percent of the direct choices with respect to informal discussion of research and 46 percent with respect to influence on the selection of problems. Four percent received 25 percent of the direct choices with respect to thesis directors.

10. In the rural sociology area, one of the nine highly productive scientists did not return the questionnaire. Two of the six highly productive scientists in the mathematics area did not return the questionnaire. Since these two individuals were very central to the area, exclusion of choices directed toward them from these computations has the effect of underestimating the amount of connectivity in the area.

In the previous chapter, it was argued that the most productive scientists are the leaders of groups of researchers. Their contacts with each other link these groups into a single network. This process has been shown in a number of ways. Following Price and Beaver (1966), the present study utilized the concept of groups of scientists linked to one another directly or indirectly by published collaborations or student-teacher ties.[11] In the rural sociology area, there were two large groups of collaborators, a few medium-size groups (five to thirteen members) and several small groups with two to four members (see Table 8). Scientists who were engaged in research in the rural sociology area were most likely to discuss current research problems with members of their own groups of collaborators (see Table 10). Almost half of the scientists named were members of the respondents' own groups. When scientists were named who were not in the same group of collaborators as the scientists naming them, the former tended to be members of large groups of collaborators. Members of medium and small groups and isolates were seldom chosen in spite of the fact that there were more members of these groups (thirty) than there were members of large groups (twenty-two) (see Table 10). The latter apparently had greater visibility in the area as a whole.

When indirect ties resulting from choices of those with whom research was discussed were examined using the Sociometric Connectedness Program, more linkage between groups of collaborators emerged, indicating that information moved among as well as within such groups. Figure 16 in the appendix, a sociogram depicting these relationships, suggests the important role that members of large groups played in tying the entire area together. The two large groups were linked to each other by reciprocal ties. Most of the medium and small groups and some of the isolates had at least one direct tie with these two groups. Some of these ties were reciprocal. For the most part, ties between medium and small groups were indirect. These groups were linked to one

11. A scientist was assigned to a particular group of collaborators if he had a published collaboration with at least one of its members or had been the student or thesis director of at least one of its members.

another indirectly through their ties with the two largest
groups (see Table 11).

These findings reflected the activities of the most productive scientists around whom these groups had formed. All of
the most productive scientists in the area belonged either to
the two large groups of collaborators or to the two largest of
the medium-size groups. Most of the members were their
collaborators or their students. As we have seen earlier a large
proportion of the choices of members of the area were directed toward them. All eight of the most productive scientists were in communication with one another directly or
indirectly about current research.

Large groups of collaborators played a similar role in the
mathematics area (see Table 9). Although there was a strong
tendency toward within-group choices with respect to informal communication and influence, members of the two
large groups also frequently chose members of the other
group.[12] Communication and influence flowed easily across
the two groups. All but one of the six highly productive
members of the area belonged to one of these groups. These
five were linked directly or indirectly by informal communication ties. This area included a large number of isolates
who looked to members of the large groups for advice and
consultation.

Crawford's analysis of sleep and dream research (1970a)
also reveals the importance of certain key individuals in the
dissemination of information throughout the field. In a group
of over two hundred researchers, 11 percent of the scientists
received 54 percent of the choices; 43 percent were never
chosen. The thirty-three most frequently mentioned scientists were surrounded individually by subgroups of scientists
who looked to them for information. They in turn communicated intensively with one another. Crawford developed the
concept of "research centers" or "areas in which scientists

12. Forty-seven percent of the direct informal communication
choices were for members of their own group and 32 percent were
for members of the other large group. Fifty-four percent of the
direct choices of members influencing their selection of problems
were for members of their own group with 29 percent going to the
other large group.

function as a unit, regardless of institutional affiliation or political boundaries of city and state. The members collaborate in research, share laboratories and are readily accessible to each other" (1970b). These centers were the foci for the activities of the scientists who were central to the communication network in the area studied. Crawford concludes that "Information generated by scientists is communicated through central persons in one research center to central persons in another center, thus cutting across major groups in the large networks of scientists. Through the central scientists, then, information may be transferred to all other scientists in the network" (1970b: 13).

Gaston (1969) found that the communication ties among 30 percent of his group of British high-energy physicists linked directly or indirectly all but one of the twenty-three British institutions where this type of research is conducted. He concluded that information circulates in a very efficient manner in that system.

Zaltman and Blau (1969) had sociometric data from 977 theoretical high-energy physicists in thirty-six countries (approximately 45 percent of the estimated world population of scientists in the area). They located a small group of thirty-two physicists who were named most often by other physicists as communication contacts outside the respondents' own institution. All of these men were linked by direct or indirect communication ties. Eighty-five percent were also frequently named by respondents in the area as being responsible for the most important research in the area. The authors label this group "a highly elite invisible college."

In areas in which a high degree of consensus regarding theory and methodology exists, the exchange of information among leaders may be formalized. An informant in high-energy physics described how prestigious members of that area attend annual conferences where they confer about the important problems to be solved in the area and make decisions about the research that needs to be done in the immediate future.

Russett (1968) detected a similar pattern from an analysis of the references appearing in the papers published in a

research area in political science. He used factor analysis of
citations to reveal the presence of twelve distinct groups
among specialists in international relations. Many of the
groups were dominated by one or two senior figures "who
had influenced groups of students and colleagues and im-
pressed their theoretical and methodological viewpoints upon
a segment of the field." He also noted considerable linkage
among these groups. Members of larger groups learned especially from the leaders of the other groups. He suggested
that members of any particular group do not have time to
examine much of the work of other groups, although to a
large extent they do keep up with what the leading figures
in the field are doing. This was shown by the fact that one
of the leaders was cited frequently by four groups other than
his own, another by three. Ten others were cited frequently
by two groups other than their own. Russett interpreted
these findings as indicating that certain figures played an
important role in linking together the potentially disparate
elements of the field.

These findings from various studies indicate clearly the
presence of an invisible college or network of productive
scientists linking separate groups of collaborators within a
research area. There is some tentative evidence that the
absence of an effective invisible college linking groups of
collaborators can inhibit the development of a field. Mc-
Grath and Altman, who analyzed the research area devoted
to research on small human groups, found that "the [typical]
study is done more or less in isolation from other small
group endeavors, in the sense that it seldom attempts to
replicate the findings, variables, or studies of others. . . .
Lack of theoretical emphasis, lack of ties with other work,
and lack of replication all are very general problems of
the field" (1966: 53–54).

They attribute this situation to the fact that researchers
are able to advance their careers most efficiently through
quantity rather than quality of publication. This in turn has
led to an emphasis upon methodological rigor rather than
substance and theory. One way to publish rapidly is to apply
"the [same] procedure, task, or pieces of equipment over and

over, introducing new variables or slight modifications of old variables" (McGrath and Altman 1966: 87). They argue that development of theory is time-consuming and is not suitable for publication in numerous segments. In the small group research area, each researcher or group of researchers has developed procedures that are not used by others in the field. As a result, research in the area has become highly differentiated, each group having its own "territory." Exchange of ideas becomes less and less necessary.

Analysis of a comprehensive list of publications in this area compiled by McGrath and Altman (1966) supported this interpretation.[13] Judging from the size and number of groups of authors linked by published collaboration, the area was highly fractionated. Seventy-two percent of the members of the area belonged to groups of collaborators with four or less members compared to 58 percent of the members of the mathematics area and 42 percent of the members of the rural sociology area. Only 18 percent of the members of the small-groups area belonged to one of three large groups (i.e., groups with more than fifteen members).[14] In the mathematics area, 42 percent of the members belonged to one of two large groups of collaborators. In the rural sociology area, 27 percent of the members belonged to one of two large groups. In the phage area, 35 percent of the members belonged to a single large group (Mullins 1968c). Price and Beaver (1966) found that 30 percent of the members of the information exchange group for oxidative phosphorylation belonged to three large groups of collaborators.

While large groups of collaborators can play important roles in the exchange of information in research areas, they may not be able to perform this role effectively when they represent a minority of the membership. They are more likely to be a minority in areas that are very large. The small-groups area had 1,452 authors compared to 221 authors in the rural sociology area, 102 authors in the mathematics area, 388 authors in the phage area, and 555 authors

13. For logistic growth of the area see Figure 11.
14. One of these groups was very large (204 members).

in the oxidative phosphorylation information exchange group.

Storer (1968) has hypothesized that each scientist defines his own set of colleagues whose work he considers to be related to his own research. Instead of social organization in research areas, he sees "custom-made reference groups." To the extent that each scientist appears to have a somewhat different definition of the membership of a research area, his concept is appropriate. Thus the existence of a group of highly productive scientists who are visible to most scientists in the area and who influence the development and direction of the field implies something more structured than individual reference groups.

The Group of Collaborators as a Source of "Solidarity"

A number of impressionistic descriptions were presented in the previous chapter, suggesting that the leaders of the large groups of collaborators recruit and socialize members and create a sense of commitment to a research area. In this section, some quantitative indications of this phenomenon will be described. As these research areas expanded in size, the influence of members of the area as compared to scientists who had not published in the area increased.[15] It was

15. For example, before 1956, about 25 percent of the theses written in the rural sociology area were directed by members of the area. From 1956 to 1966, close to two-thirds of the theses in the area were directed by members of the area. The numbers of scientists mentioning only outsiders as influences upon their selection of research problems in the area decreased from 38 percent among scientists entering before 1951 to 9 percent among scientists entering after 1956. (The date of a scientist's first publication in the area was used as the date of his entry into the area.) The proportion of collaborative publications increased from 22 percent before 1951 to 30 percent between 1951 to 1955, 41 percent between 1956 to 1960 and 37 percent after 1960. These last two distributions suggest that the field was somewhat more integrated from 1956 to 1960 (the second half decade of rapid exponential growth) than at other periods. A slight decline in integration parallels the shift from exponential to linear growth.

In the mathematics area, the proportion of collaborative publications increased over time in this area as in the rural sociology area, although it never comprised more than 18 percent of the publications in any period. Direction of theses by members of the area increased

possible to observe the emergence of certain individuals who contributed numerous publications to the area and who exerted considerable influence upon its development. Most of these scientists published their first papers in their fields either before or during the period of exponential growth.[16] These individuals maintained their interest in these areas for considerable periods. In the rural sociology area, the average career of the relatively unproductive members lasted for three years, the average career of the moderately productive members for seven years, and of highly productive members for eleven years.[17] In the mathematics area, the comparable figures were three-and-a-half years for less productive members, eight-and-a-half for moderately productive members, and sixteen-and-a-half for highly productive members. Thus the highly productive scientists were able to have a substantial impact upon their research areas in terms of setting the norms for research and of maintaining the cohesion of the area. They exerted this influence by surrounding themselves with students and collaborators. The period of exponential growth in both these areas was characterized by the emergence of large groups of collaborators to which most of the highly productive scientists belonged.

During the period of exponential growth, the highly productive scientists played major roles in the direction of

---

as the field grew, as did the influence of members of the area upon the selection of problems compared to that of nonmembers (see chap. 6). Thirty-three percent of the theses were directed by members before 1950. Eighty-one percent of the theses were directed by members in 1960–68.

16. It is obvious that scientists who enter a field earlier in its development have more time in which to be productive. In this study, productivity has been measured in terms of the total number of publications produced by a scientist. Productivity could also be measured in terms of the number of publications produced by a scientist in relation to the number of years he had been working in the field. This measure of productivity estimates it in relation to time only. The absolute number of publications per scientist is an indication of his impact upon the field. The latter measure is correlated with influence upon other scientists as measured by their sociometric choices. In this sense, the scientists who have the most impact upon the field enter relatively early in its development.

17. Length of career was determined on the basis of the dates of the author's first and last publications in the research area.

theses and as sources of influence upon the selection of research problems. After 1956, 38 percent of the theses in the rural sociology area were directed by these eight scientists (none of the theses written in the area before that period had been directed by them). Twenty-seven percent of the theses in the area were directed by the remaining 139 respondents. Similarly, the proportion of scientists in the area mentioning the influence of the highly productive scientists upon their selection of research problems increased from 25 percent among scientists entering prior to 1951 to 39 percent among scientists entering after that date, more than twice as high as the proportion naming the other members of the field. More than 50 percent of the total number of choices of this type by scientists entering the field after 1950 were for these highly productive scientists.

In the mathematics area, after 1955, almost one-third of the theses in the area were directed by one of the six highly productive members. The proportion of scientists in the group mentioning highly productive scientists as having influenced their selection of research problems in the area rose from 22 percent during the first decade of exponential growth to 39 percent during the second decade. Forty-four percent of the total number of choices of this type for members of the area were for the six highly productive members.

Groups of collaborators that do not have aggressive leadership are unlikely to endure. This type of leadership is reflected in the size of the group. In both areas, there was a strong relationship between the size of a group and the number of years it lasted (see Tables 8 and 9). Sixty-one percent of the groups of collaborators in the rural sociology area had lasted less than five years. New members were most likely to be available for the expansion of groups of collaborators when graduate students were being trained in the discipline in the same academic setting. There was some indication that groups of collaborators whose leaders were located in universities that trained sizable numbers of sociologists were larger (Cartter 1964).

Foreign settings appeared to be unsuitable for the expansion of these groups. Small groups of collaborators had higher proportions of foreign members than large or medium-size groups. In both areas, almost half the isolates were residing outside the United States. Lack of students and frequent contacts with colleagues, which constitute the resources for expansion of such groups, are probably the explanation for this situation in many countries.

Differences in the type of communication that occurs may be an important aspect of the research environment provided by the large groups of collaborators. Communication among scientists appears to take at least two forms, which are revealed by reciprocal and nonreciprocal choices. Much scientific communication is purely consultative, one scientist asking the advice of another. This is reflected in a high proportion of unreciprocated choices.[18] Reciprocated choices would appear to reflect the occurrence of an exchange of ideas. Such ties were more likely than nonreciprocal ties to be rated by the scientists concerned as being very important to their ongoing research.[19] While less than one-third of the direct choices made by members of the rural sociology area were reciprocated, such choices that did occur were more likely to have been made by members of large groups. Fifty-eight percent of the reciprocated choices were made by members of large groups. For large groups, the number of reciprocated choices per member was 0.68; for smaller groups and isolates, it was 0.36. Apparently, large groups of collaborators provided more opportunities for the exchange of ideas.

The situation in the mathematics area was different from that of the rural sociology area in that, almost half (40 percent) of the 102 members of the area were isolates (i.e., had no collaborative or student-teacher ties).[20] Two large

18. Among those doing research in each area, the proportion of informal communication choices which were reciprocal was small (31 percent in the rural sociology area, 16 percent in the mathematics area). Crawford (1970a: 40) also found a small proportion of reciprocated choices: 33 percent.

19. Sixty-nine percent of the reciprocated ties in the rural sociology area were rated as "very important" (i.e., given a rating of four or five out of five) for their research by respondents; only 41 percent of the nonreciprocated ties were so rated ($N = 83$).

20. There were also nine two-person groups.

groups of collaborators with 24 and 19 members, respectively, represented 42 percent of the members of the area (see Table 9).[21] Hagstrom (1965: 228) has observed that many mathematicians have difficulty identifying their colleagues; the audience to which they address their research is unknown to them. As a result, they find it difficult to evaluate either their own research or the importance of their areas in relation to other research areas. They tend to label the work of others in their own areas as trivial. Hagstrom considers this situation anomic. He defines anomie as the absence of opportunities to receive recognition. According to Hagstrom, this state of affairs, which is demoralizing for a scientist, tends to occur in mathematics because in that discipline researchers tend to be highly specialized but their specialties are not clearly defined. One consequence of this may be that many mathematicians enter fields in which they have neither student-teacher nor collaboration ties. They and their contributions remain marginal to the area and receive little recognition within it. These ideas can be tested by examining the characteristics of the communication and influence networks in the mathematics area. Were those who did not belong to groups of collaborators less likely to be linked to the social networks of the area than those who were members? If so, they might be relatively unaware of the norms defining important work in the field since the latter are set by members of large groups. Consequently, their work would have less influence upon other members of the field than that produced by members of large groups.

One of the large groups of collaborators in the mathematics area was similar to the large groups in the rural sociology area. It was dominated by a highly productive mathematician who had trained or collaborated with half of the other members of the group. In turn, a few of his

21. One mathematician who had worked in the area during the early years of the twentieth century had taught one member of both groups. Since this was the only common link between the two groups, they were considered as separate entities. This individual was not included in calculating the duration of these groups.

students and collaborators (one of whom was highly productive) had acquired students and collaborators of their own, producing smaller subgroups within the group. The other group of collaborators was based principally upon collaboration rather than student-teacher ties.[22] Three highly productive members had collaborated with each other and with several others. These three men provided some focus for the group that grew by adding collaborators instead of students. This was surprising since all three men were located in universities where graduate students were available.

Both isolates and members of two-person groups were much less likely to be named as persons with whom members of the area discussed their current research than were members of the large groups of collaborators (see Table 12).[23] They also made fewer such choices. Nor were most isolates or members of two-person groups linked indirectly to members of the area through choices of other individuals (see Table 13).

While this would suggest that these mathematicians were less integrated into the social networks of the area than were members of large groups of collaborators, it is possible that the quantity of ties is less important than having at least one. Isolates were even more likely than members of large groups to report that informal communication was very important to their research.[24] To what extent were the isolates and pairs aware of at least one other person who was working in the area? Isolates were less likely either to report informal communication with or to be mentioned in this connection by even one member of the area than were members of large

22. There were fourteen student-teacher relationships in the first group compared to six in the second group.
23. When tables 12 and 13 were presented in the same way as tables 10 and 11, the findings were similar.
24. Forty-eight percent of the members of large groups rated their informal communication contacts as very important to their research; 57 percent of the isolates did so. Only 13 percent of the members of two-person groups did so ($N = 44$). Forty-four of the sixty-four respondents in the mathematics area were continuing to do research in the area.

groups.[25] Fifty percent of the isolates, however, were aware of at least one other person working in the area as indicated by their own choices. Others were less likely to be aware of the isolates since only 21 percent of the isolates were named by others. The fact that they were as likely as members of large groups to send preprints of their work suggests that they were striving to increase their visibility in the area.[26]

The American members of the area held frequent conferences at different sites. In a recent four-year period, there were at least six conferences devoted exclusively to research problems in this area. Conferences could be lengthy, lasting anywhere from two days to two weeks. Isolates and members of two-person groups attended such conferences as frequently as members of large groups.[27] At various times, members of the area had gathered at one institution for one or more months. Visiting appointments that brought two or more members together at one university for the academic year were also frequent. Again, isolates were as likely as members of large groups to have participated in these types of activities.

These findings appear inconsistent. If isolates and members of two-person groups attended conferences and circulated preprints as frequently as members of large groups, why were they less frequently named as informal communi-

25. Members of the various types of subgroups chose and were chosen *at least once* in the following proportions (informal communication choices only):

	Large Groups	Pairs	Isolates
Percent chosen	60	40	21
Percent choosing	80	60	50
Number	(25)	(5)	(14)

Goodman and Kruskal's gamma is .34 and .25 respectively.

26. Twenty percent of the members of large groups, 60 percent of the members of two-person groups, and only 7 percent of the isolates sent no preprints of their work to other members of the area ($N = 44$). These figures are substantially lower than those reported by Hagstrom (1970: 89) based on a sample of mathematicians drawn from *American Men of Science*. He reports that 54 percent of his sample did not send out preprints of their publications.

27. Sixty-four percent of the members of large groups, 8 percent of the members of two-person groups, and 78 percent of the isolates reported attending conferences in their area at least once or twice a year ($N = 44$).

cation contacts? Part of the answer may lie in the ecology of the area.[28] An informant described one aspect of this: "Boston is the largest mathematical research center in the world. Chicago is another center. It is important to be in such a center if you want to be part of the network of communication."

One-third of the members of the large groups were located as visitors or as permanent faculty in either of these two centers at the time the study was done. Fifty percent had been located in such places if previous visiting appointments were included in the tally. None of the members of two-person groups and only 14 percent of the isolates had ever been located in those two centers. In addition, these mathematicians were much more likely to be residing abroad than members of the large groups.[29]

While this area does not exhibit the acute social isolation that Hagstrom described, it is clear that some members were less involved in the communication network than others. In turn, their work appeared to have had less influence upon the research of others in the area. Members of two-person groups and isolates were named less frequently than members of the large groups as having influenced the selection of research problems by members of the area. They did not seem to have had such an influence through third parties. These data suggest that research by pairs and isolates had less influence upon the development of the field than that of members of large groups. They were, however, aware of the research of other members of the area. Isolates and members of two-person groups were as likely to name at least one member of the area as having influenced their selection of problems as were members of large groups.[30]

28. Allen (1969a) found that the ecology of a research organization affected informal communication patterns. The greater the distance between offices, the less frequent the contacts between their occupants.

29. Sixty percent of the members of two-person groups and 43 percent of the isolates were residing abroad, compared to 17 percent of the members of the large groups ($N = 44$) (Goodman and Kruskal's gamma is .34).

30. Pairs and isolates who together comprised 58 percent of the members of the area received 31 percent of the direct choices and

According to an informant in the area, the work of the isolates was uniformly of low quality, and this factor explained their relative lack of influence in the area. He claimed that those who did produce high-quality work were immediately invited to join the larger groups. It appeared, however, that standards for evaluating the quality of research were set by members of the large groups of collaborators and that isolates and members of small groups did not conform to them. Were those who did conform indeed admitted to those groups? This did not seem to be the case, since few members of the large groups had begun their careers as isolates. Publications by pairs or isolates that were cited frequently were published early in the history of the area when standards for research in the area were still being formed. Later it appeared that members learned these standards by being socialized from the beginning of their careers in large groups of collaborators.[31]

International Scientific Communities

There is also some evidence that social circles transcend national boundaries. Ben-David (1968) has argued that the scientific community has been international since its emergence in the seventeenth century. And Zaltman (1968) has shown that the field that is presently called high-energy physics and that evolved from atomic and nuclear physics has an international communication network. This network has three major subsystems, consisting of the United States, the European countries that participate in the international high-energy physics laboratory (Conseil Européen pour la Recherche Nucléaire, C.E.R.N.), and Japan. Approximately

---

37 percent of the indirect choices representing influences on the selection of research problems. Seventy-nine percent of the isolates named at least one member of the area as an influence upon the selection of research problems. Seventy-nine percent of the members of large groups and 64 percent of the two-person groups did so ($N = 64$).

31. The correlation (Goodman and Kruskal's gamma) between collaboration group affiliation of senior author (group, pair, or isolate) and number of senior authors citing his publication was .06 during the period before 1950, .33 between 1950–59, and .29 between 1960–68.

88 percent of the members of the field are working in one of these three regions.[32] Zaltman found that Japan was relatively isolated from the flow of informal communications, but that Japanese contributions to research were acknowledged as frequently as contributions from the other two regions.

In the mathematics area, the structure of the international network was similar to that described by Zaltman in physics. Mathematicians from the United States and three European countries played important roles in this network. Non-Western mathematicians did not appear to have much influence. Papers published in non-Western journals were much less likely to be cited even once than papers published in American and European journals. In the rural sociology area, no other country or group of countries played a role comparable to that of the United States. In all three of these areas, the United States alone had more members than the largest concentration of members in any other country; this disparity was greatest in the rural sociology area. Almost one-third of the non-American members of that area had less than three colleagues in their own countries, compared to 14 percent of the non-American mathematicians and 2 percent of the high-energy physicists employed in countries other than the United States (Zaltman and Köhler 1970: 4). Only 18 percent of the members of the rural sociology area who had produced more than three papers were not American; the comparable figure for the mathematics area was 54 percent.

It appears that communication networks in research areas are effective in linking scientists from different countries, but that scientists in some regions are less involved in these networks, and, consequently, their work is less visible to their colleagues in other countries. Disciplines differ considerably in the degree of participation by different countries in the international scientific community. Opportunities for developing a long-term commitment to an area with concomitant productivity appear to be related to the availability of a "critical mass" of colleagues in the same geographical region.

32. Since Zaltman's survey did not receive any replies from physicists in the U.S.S.R. or the People's Republic of China (Communist China), these groups were not included in his analysis.

4
Social Organization and the Diffusion of Ideas

IF SOMETHING COMPARABLE TO A PARADIGM INFLUENCES the development of research, an analysis of the literature of a research area ought to reveal its existence. If this model is correct, important ideas initiate the growth of a research area. The importance of research is determined according to standards set by members of the area. A period of "normal science" follows, during which subsequent research builds upon these innovations and remains closely related to them. If paradigms are gradually exhausted, one would expect to find, over time, a decline in the proportion of important innovations. Many of the later innovations are relatively minor, examining the implications of the earlier work. When the ideas have been thoroughly explored, the number of new recruits to the field decreases. Alternatively, anomalies that the original paradigm cannot explain lead to the development of a new paradigm that eventually replaces the earlier one.

In order to explain the growth and development of knowledge in a research area, a model must account for the diffusion of ideas in that area. It ought to be possible to detect, first, which scientists are setting the standards for evaluating research in an area and, second, a "contagion" process that transmits important ideas to a steadily increasing population of "adopters" who explore their implications. The development of a paradigm is both an intellectual and a social process in which ideas are evaluated and norms for subsequent research are set. Analysis of the social organization of research areas has shown that a few scientists in each area played very important roles in recruiting and influencing other members. This suggests that consensus concerning a paradigm for an area may emerge among a small group of scientists who then transmit it to many others.

Innovation and Scientific Growth

It is hypothesized that research areas develop as a result of important innovations that appear prior to the period of exponential growth. Later work explores and develops the implications of these ideas. In the rural sociology area, a significant proportion of the innovative work in the area had already been done by the time the field began to acquire a significant number of new members (see Table 16). During the first decade (1941–50), the number of innovations (i.e., new variables utilized in empirical studies) increased rapidly, although growth in the number of publications and of new authors entering the area was very slow.[1] Almost one-third of the innovations appeared during this ten-year period. This suggests the role of something analogous to a paradigm. These early studies attracted considerable attention outside the area (Rogers 1962: 32).

The rate of innovation declined during later stages of growth in the rural sociology area. During the period when the rate of growth in the area was beginning to taper off, only 17 percent of the innovations were produced in spite of the fact that the field was very productive at that time. During that period, 45 percent of the empirical publications

1. For further information about the data see Chap. 1 and Rogers et al. (1967). Publications included journal articles, papers read at professional meetings, papers published in agricultural extension bulletins, and theses. Information was not available for 13 of the 342 publications containing empirical data. Ten of the 213 authors who had produced empirical publications were the authors of these 13 publications. Sixty-one theoretical publications were also included in the bibliography but were not included in the study since they had not been content-analyzed by Rogers et al. (1967).

appeared, and 44 percent of the authors entered the field. This finding fits Holton's hypothesis that "the number of active researchers will reach a maximum when a large part of the presumed total of interesting ideas has already been discovered" (1962: 391).

There was a strong correlation between the date of publication of an innovation and the number of adopters. Innovations produced during the earlier periods of activity in the field were used by many more members of the area than innovations produced in later periods, in spite of the fact that the majority of the members of the area published for the first time in the area during the last decade covered by the study (see Table 17).

There was also some evidence that the early period in the development of the mathematics area was a seminal one. While the numbers of publications and of authors entering the area increased over time, the percentage of publications not citing previous publications declined. Using the latter as a measure of the amount of innovation per period (Garfield et al. 1964: 32), results were obtained that were similar to those in the rural sociology area. Over one-third of the papers that did not cite previous papers in the area appeared before the numbers of publications and authors had increased substantially (see Table 22). Other ways of measuring this type of innovativeness revealed the same trend (proportion of nonciting publications per period, ratio of nonciting publication to number of new authors per period).

When innovativeness was measured in terms of the number of times a publication was cited, the results were less clear. In the rural sociology area, the earliest innovations were most frequently utilized. In the mathematics area, the earlier publications were cited by more senior authors than later ones, but the correlation between date of "innovation" and frequency of usage was much lower.[2] In part, this can be explained in terms of the history of the area. According to informants, the field had been active in the nineteenth century and largely dormant since then until the fifties when new work attracted considerable interest and the field began

---

2. Pearson $r = 0.12$ (computed from grouped data).

to expand both in terms of numbers of publications and authors. On the other hand, authors entering the field before 1940 were much more likely to be frequently cited than authors who entered at a later date (see Table 23).

Of the ten most frequently cited publications (those cited more than ten times), two appeared before 1940 and four each in the decades 1950–59 and 1960–68. Thus, according to this criterion, "major" innovations continued to appear throughout the history of the field. These publications, however, were linked through their citations to early publications in the area. The model suggests that we should find a reverse pyramid with a few publications at the base and a large number of publications linked directly or indirectly to them at the top.[3] Six of the ten most frequently cited publications in the area (those cited by more than ten senior authors) cited at least one publication that had appeared before 1940. Two others appeared during this period. Of the thirty-seven publications that were cited by six to ten authors, six were published before 1940 and fourteen others cited at least one such publication. Altogether, of the forty-seven publications cited more than five times, 60 percent were either published before 1940 or cited publications from that period.

Finally, later work relied heavily on five "classical" authors whose publications appeared at the turn of the century and were not included in the bibliography.[4] A paper reviewing the area in the early sixties (Gorenstein 1964) made several references to three of these authors. Of the forty-seven most frequently cited publications in the area (cited more than five times), 66 percent cited at least one of these five authors. Eighty percent of these most frequently cited publications fell into one or more of the following categories:

3. The analysis by Garfield, Sher, and Torpie (1964) of citations contained in publications dealing with diverse aspects of the genetic code also examines chains of influence linking publications. Their evidence does not fit this model, probably because they do not confine their attention to a single research area. The publications with which they were concerned dealt with "nucleic acid chemistry, protein chemistry, genetics, microbiology, or pertinent combinations of these disciplines" (p. 3).

4. They were W. Burnside, H. F. Blichfeldt, L. E. Dickson, G. Frobenius, and C. Jordan.

(1) published before 1940; (2) cited publications in the area that were published before 1940; or (3) cited one or more of the five classical authors.

It appeared that the foundations of the mathematics area were laid before 1940, approximately ten years before the onset of exponential growth in the area. This does not mean that significant innovations did not appear in the area after that time. In this field, some very important papers appeared in the late fifties and early sixties (Gorenstein 1964). It is simply being argued here that these papers built on and extended earlier work. This is less inevitable than it seems. If it is true that early researchers in a field "define" the field and develop some of the general methods for attacking problems in it, it might appear that anything that comes later "builds on" what these papers have done. Later papers that did not build on early work were generally ignored. More than half of all papers published after 1940 that did not cite previous work failed to generate any new research (as evidenced by citations). This was not true in the early stages of development of the field. Only one of the twelve papers published before 1940 that did not cite other papers in the area failed to be cited. It is possible that after a certain point in its development a field is not susceptible to new directions.

The Diffusion of Ideas: A "Contagion" Process

Several studies have shown that most publications produced in a field are seldom cited, while a few are cited by many authors (J. Cole 1970; Kessler and Heart 1962; Price 1965a; Weinstein n.d.). This pattern appeared in both the areas being examined here. In the rural sociology area, 21 percent of the innovations were used by only one individual, the innovator. Fifty-eight percent were used by five or fewer members of the area. Forty percent of the publications in the mathematics area were never cited.[5] That these publica-

5. Kessler and Heart (1962) in their study of papers published in the *Physical Review* found that only 10 percent were never cited. The percentage of papers never cited in the mathematics area is higher due to the fact that self-citations were excluded.

tions did belong to that area is largely confirmed by the fact that only 24 percent of them did not cite other publications in the area. After 1940 the proportion of uncited papers remained fairly constant throughout the development of the field.

It is often argued that frequency of citation reflects the quality of the work being cited (J. Cole 1970; S. Cole 1970; Cole and Cole 1967 and 1968). Within a research area, research that is frequently cited conforms to standards that have been set by members of the research area. If this is the case, some individuals can be expected to play a greater role in this process than others. In turn, we would expect to find evidence of personal influence upon the diffusion of ideas from those publications that were being used to set the standards for research in the area. The term "standards" is being used here to mean not standards regarding the evaluation of completed research but standards concerning the types of hypotheses to be examined and the types of data to be collected.

The pattern of dissemination of innovations that were frequently used in the rural sociology area suggested that personal influence was having an effect. The size of the area prohibited an enormous expansion in usage, but the pattern of continual increase in adoption indicated the presence of a contagion process. This hypothesis was confirmed by plotting the cumulative number of adoptions over time for the most widely used innovations (see Fig. 13). These curves are similar to the logistic curve that previous studies have interpreted as indicating a process of diffusion as a result of personal influence (Coleman et al. 1966; Dodd 1955).

The effect of personal influence upon diffusion can also be seen in the fact that, after 1956, when the number of scientists entering the field increased considerably, innovations that had been produced and most frequently used before that date continued to be widely used afterward (see Table 18). Those that had been seldom used before that date were seldom used afterward. In the same group of innovations, the number of adoptions between 1961 and 1966 was related to the number of adoptions between 1956 and 1960 (see

Table 19). Thus those that were frequently used between 1956 and 1960 were frequently used between 1961 and 1966.

A similar process could be detected in the mathematics area. When the cumulative number of adopters of the ten most frequently cited publications was plotted by year, a logistic curve was generated (see Fig. 14). The number of senior authors citing a publication in a later time period was related to the number of senior authors who had cited it in an earlier period (see Table 24). Crawford's data (1970a: 48) in the area of sleep and dream research also exhibited this pattern. Papers that were cited most during the years immediately after publication continued to be cited most frequently in subsequent years.[6] For reasons that will be discussed later in the chapter, the diffusion of ideas produced in subsequent stages of the development of the first two areas did not exhibit such a "contagion effect."

The Role of Opinion Leaders

Studies of voting decisions and of purchasing habits have indicated that the transferral of information from mass media to the public is mediated by "opinion leaders" who absorb more information on specific topics than the average person and in turn communicate it to others (Katz 1960; Katz and Lazarsfeld 1955). Coleman, Katz, and Menzel (1966) in their study of drug adoption by physicians found that early adopters of a drug read more medical journals and were in contact with more members of the profession than physicians who were late adopters. Information of a technical nature appears to be transmitted to a large public through personal contacts.

It appeared that certain ideas were more influential than others in the development of the two research areas and that personal influence played a role in disseminating them. Who were the "opinion leaders" in these areas? Did they promote their own work or that of others? In the rural

6. Crawford's findings (1970a: 48) were based upon a random sample of fifty papers published in the area of sleep and dream research.

sociology area, innovations that were adopted by the most productive scientists were more widely accepted by other members of the field than innovations that they did not adopt(see Table 20). In the mathematics area, also, publications cited by one of the six highly productive members were more frequently cited than publications that they did not cite (see Table 25).[7]

The relationship between productivity and the production of ideas in these areas was complex. The distribution of "innovativeness" in the rural sociology area was similar to the Lotka distribution of productivity. In both cases, a few scientists in a population were responsible for a high proportion of the total output while a high proportion of the members made few contributions (Price 1963, chap. 2). Fifty-one percent of the 203 authors in the area had been associated as either a senior or junior author with a publication in which an innovation had appeared.[8] Only 27 percent of the authors had produced more than one such publication, and only 6 percent had produced more than five. Thus, while 49 percent of the members of the area had not produced any innovations, six members (3 percent) had participated in the production of 52 percent of the innovations.

Innovators tended to be scientists who had produced several publications in the area. Sixty-four percent of the innovations had been produced by senior authors who had published four or more papers in the area, although these authors represented only 15 percent of the total group of authors. Moderately productive scientists, however, produced more innovations than highly productive ones. This was due in part to the fact that many innovations were produced by scientists who entered the field very early and did

7. This effect is not due to the fact that these authors were producing more papers in which they cited other papers since the dependent variable here is the number of senior authors citing, not the number of papers containing citations.

8. In 10 percent of the cases, an innovation appeared for the first time in two or more publications in the same year. For the purpose of assessing the characteristics of innovators, such publications were assigned to different categories on a fractional basis.

not continue to work in it. For example, fifteen of the eighteen most widely used innovations (those with more than twenty users) were published by two moderately productive members of the area who entered the field during its second year of existence. These two scientists together were associated with 21 percent of all the innovations in the area.[9]

It appeared that some members of the area may not have realized the original source of some of the innovations. Data from the questionnaires showed that the highly productive scientists were more likely than the moderately productive scientists with the same number of innovations to be credited with having influenced members of the field.[10] Something analogous to what has been called in other contexts the "two-step flow of communication" (Katz 1960) seems to have occurred, with the highly productive scientists acting as intermediaries in transmitting the innovations made by the first scientists in the area to those who entered the field at a later date. This is illustrated in Figure 18, which shows that members of large groups of collaborators named directly as sources of influence upon their selection of research problems members of two medium-size groups that included four of the eight scientists who had published in the area during its initial period of irregular growth. Members of one-third of the medium and small groups and isolates were linked to those groups through their choices of members of the large groups. Later the likelihood decreased that members of medium-size groups would have an influence upon the direction of research in the field.

In the mathematics area, also, the correlation between productivity and frequency of citation (the measure of innovativeness being used in this area) was low (Pearson

9. Crawford (1970a: 60) also found that two early important contributors in the area of sleep and dream research were heavily cited during a period when they produced only one paper each.

10. Only twelve authors had produced more than five innovations. All of the highly productive members of this group of twelve were mentioned more than five times by respondents as a source of influence upon their selection of research problems in the area. Only 33 percent of the moderately productive innovators were mentioned more than five times by respondents in answer to the same query.

$r = 0.20$). This low correlation was the result of changes over time in the characteristics of the members who produced the most cited publications. Among publications appearing before 1950, publications by the least productive authors were somewhat more likely to be cited than those by the most productive authors (see Table 27). In the fifties and particularly in the sixties, the reverse was true. Publications by the most productive authors were much more likely to be cited, and were cited more frequently, than those by less productive authors.[11] It appeared that less productive authors were more likely to have an impact upon the field in the early stages of its development.

The influence of the most productive members of the area as adopters was strongest with respect to publications appearing early in the development of the area (see Table 26). Eighty-three percent of the most productive members and 80 percent of the moderately productive members cited one or more authors who published before 1940. Thus, while their influence as authors increased over time, their influence as adopters decreased. This study suggests that opinion leaders are the most productive scientists in an area. They stimulate growth of an area by their adoption of innovations produced early in its history. These ideas are widely used by members of the area, but gradually new work by the most productive scientists based on these ideas becomes more influential in the area.[12]

Another study (Kessler and Heart 1962) has shown that if publications are not cited soon after their appearance, they are unlikely ever to be cited. Similar findings occurred in both these areas. Of the innovations that were adopted in the rural sociology area, only 28 percent were adopted more than five years after their first appearance. In the mathe-

11. When author's affiliation or lack of affiliation with a group of collaborators was used as the independent variable, the findings were very similar. Members of small groups and isolates were more likely to have an influence upon the field early in its history.

12. In the rural sociology area, the influence of the most productive scientists was strongest both as adopters and as authors in the early stages of development of the area. During the latter stages, no innovations were widely adopted (see Table 17).

matics area, 47 percent of all the publications (73 percent
of those cited) were first cited within five years of publication. This suggests that papers are assessed soon after publication and are unlikely to be reassessed.

Diffusion and Scientific Growth
The factors affecting the diffusion of innovations in science
appear to be similar to the factors affecting the diffusion of
other types of innovations. There is some evidence that the
diffusion of a scientific innovation is a fashion-like process
in which influence is transmitted through steadily expanding
networks of scientists. Thus it is plausible to view science as
an enormous cluster of innovations, of which the most successful are diffused by means of a contagion process that
produces a logistic curve in all facets of scientific activity.
Behind the seemingly impersonal structure of scientific
knowledge, there is a vast interpersonal network that screens
new ideas in terms of a central theme or paradigm, permitting some a wide audience and consigning many to oblivion.

There has been some disagreement over the implications
of the fact that most scientific innovations fail to diffuse
widely. Weinstein has suggested that since most scientific
information is obsolescent from the moment it appears and
is never incorporated into the body of scientific knowledge,
it is not meaningful to include these innovations when
measuring scientific growth. If these innovations were excluded, the growth of science would appear to be much more
gradual than previous studies have suggested.

J. Cole (1970), who examined the characteristics of citing and cited scientists in physics, found that members of
the most prestigious departments cite scientists in prestigious
departments more than they cite scientists in less prestigious
departments. Publications by scientists at less prestigious departments are cited by scientists at such departments and
by scientists at prestigious departments at about the same
rate. Cole concludes that a relatively small number of physicists produce the research that becomes the basis for future
scientific discoveries. He states (p. 400): "Clearly most of
the published work in even an outstanding journal makes

little impact on the development of science." He interprets his findings to mean that the size of the scientific community could be greatly reduced while still maintaining the same rate of advance in knowledge. Whether or not this interpretation should be accepted depends upon how one evaluates the significance of the seldom-cited paper. If one interprets frequency of citation as a measure of the quality of a paper, as Cole has done, then his interpretation is justified. Alternatively, it can be argued that there are differences in the types of innovations that are produced at different stages of the development of a research area. Back's (1962) model of the ways in which scientists utilize the scientific literature is relevant here. He inferred that the first generation in a research area makes a broad, sweeping search of the literature while the second generation has a more specialized orientation toward it. One could also argue that the ideas produced by the first generation are broad and sweeping while those produced by the second generation are more specialized. If this is the case, it is not surprising that the work of the first generation is more frequently cited than the work of the second generation. Later and less cited papers in an area are exploring the implications of earlier work, eliminating false leads, and providing evidence for hypotheses. In this sense, these papers are not useless. They may be read and absorbed by the working scientist, but they do not contain the sort of material that itself generates new studies.[13]

Concomitantly, other changes take place in the research area that make it less likely that new ideas will be widely used. The fact that there are more people working in the area may dampen the personal influence process that produces diffusion. The increasing number of new papers being produced may make it more difficult to spot new ideas. Dodd

13. Gaston's finding (1970) that high-energy physicists doing theoretical work obtain more recognition with fewer publications than high-energy physicists doing empirical work can probably be explained in this manner. The theoretical physicists are producing the seminal ideas that are tested by the empirical physicists who in turn cite the theoreticians frequently but are not cited frequently themselves.

(1955), in a study of the diffusion of information appearing in leaflets dropped on towns from airplanes, found that when the size of his sample increased, the proportion of adoptions decreased.

There is some indication that scientific theories have an effect comparable to that of a "mental set" influencing the scientist's perception of phenomena and his selection of relevant materials from the scientific literature (Kuhn 1962). Engineers have been found to become committed to technical approaches and to disregard information that contradicts the validity of their assumptions (Allen 1966). Among scientists who conduct basic research, paradigms produce strong mental sets. Shared "coding schemes" or common ways of ordering the materials of the field must appear if communication is to be facilitated. With time, increasingly sophisticated theoretical models may develop. All of these make it difficult for the scientist to perceive the possibility of other types of innovations.

Finally, as the group grows in size, subgroups increasingly may provide boundaries beyond which ideas do not disseminate. In other words, the field may become increasingly fragmented by "schools" representing different orientations toward the area.

Diffusion of Ideas Produced During Periods of Exponential and Linear Growth

In the rural sociology area, the period of exponential growth began about 1950 and continued for approximately ten years. In general, innovations introduced in the area after 1956 were less likely to produce expanding networks of adopters than were innovations produced between 1941 and 1955. While 48 percent of the innovations in the area were produced after 1956, none of them obtained a large number of adopters (see Table 17). There was no relationship between the number of adopters of these innovations between 1956 and 1960 and the number of adopters of these innovations in the later period. Innovations produced after 1956 had fewer users between 1956 and 1966 than did innovations produced before 1956 (see Table 19).

The influence of the highly productive scientists was greater with respect to innovations produced before 1956 than afterward. Innovations produced before 1956 and adopted by these scientists were more widely diffused after 1956 than innovations that they did not adopt. Innovations produced after 1956 and adopted by these scientists were used only slightly more often than those that they did not adopt.[14] None of the innovations produced after 1956 achieved really wide usage in spite of the fact that 84 percent of the empirical publications in the field appeared during this period.

It is possible that later innovations were less visible to members of the area than earlier innovations. The ratio of innovations to papers was high before 1956 but declined sharply thereafter (see Table 16). During the latter period, the number of papers increased more rapidly than the number of innovations. When there were many more papers than innovations being produced, it may have been increasingly difficult for members of the area to spot innovations.

Ninety-six percent of the theoretical publications in the rural sociology area appeared after 1956, which indicates that members of the field began at that time to experience a need to create some order out of the mass of empirical findings that had been accumulating. Since none of the innovations produced in the later period achieved wide usage, it appeared that no theory emerged that was effective in pointing out the relevance of some variables rather than others.

It appeared that members of the field concentrated their attention upon a certain type of problem. Almost half (48 percent) of the innovations in the area belonged to a single category of variables. The remaining variables were spread fairly evenly among the other seven categories.[15] As a result, in the later periods of activity in the area, they may have been

14. Seventy percent of the pre-1956 innovations that the most productive scientists adopted were adopted by more than five members of the area compared to 6 percent of those that they did not adopt. The comparable figures for the later period were 30 percent and 18 percent.

15. The categories are described in Rogers et al. (1967).

producing innovations that were increasingly specialized and less original. McGrath and Altman (1966: 53) in their study of small-groups research report a similar concentration of interest within that area. They classified variables appearing in studies in the area into 92 categories. One of these variables appeared in almost 30 percent of the studies while nearly one-third of these variables appeared in only a single study and over two-thirds in seven or fewer studies. Further indication that members of the rural sociology area became less innovative over time can be seen in the fact that the ratio of innovations to authors declined steadily.

Previous studies suggest that there are tendencies for members of groups of collaborators to build upon their own work rather than that of members of other groups. For example, in the phage area where there was only one group of collaborators, the group cited its own work more heavily than that of nonmembers (Mullins 1968c). Price and Beaver (1966) found some indication that the level of self-citation among the groups of collaborators studying oxidative phosphorylation was considerably higher than that for science as a whole (25 percent compared to 10 percent). Russett (1968) in his study of political scientists specializing in international relations found a strong tendency for members of subgroups within the area to cite other members of their own group.

The members of the two largest groups of collaborators in the rural sociology area also became less likely to adopt each other's innovations as the field developed. Over time, the larger groups of collaborators appeared to become more autonomous and perhaps more committed to their own point of view. Innovations produced before 1956 were much more widely diffused among groups of collaborators than were those produced in the later period (see Table 21). Among the innovations produced by members of the two large groups of collaborators before 1956, 85 percent were used by members of both groups. This was true of only 29 percent of the innovations that the two groups produced in the later period ($N = 74$ for both periods). This indicates that, over time, the two groups became less receptive to each other's innovations. They also came to differ in their selection of innovations produced by the smaller groups of collaborators and the isolates. Of the innovations produced by the smaller groups and isolates before 1956, 52 percent were adopted by both large groups. Of the innovations produced by these groups and the isolates in the latter period, only 31 percent were adopted by both groups ($N = 127$). These data suggest increasing fragmentation within the area. It appeared that the large groups of collaborators were beginning to resemble schools in the traditional sense of the word.

At the time the mathematics area was studied, the period of exponential growth had ceased. Exponential growth began about 1950 and continued until about 1965. Again, the contagion process in which early adopters influenced later adopters was weaker in the later stages of the development of the area. While the number of senior authors citing a publication in a later time period was related to the number of senior authors citing it in an earlier period, this correlation was higher for publications that appeared before and during the first five years of exponential growth than for publications that appeared during the second five years of exponential growth (see Table 24).

According to an informant, members of the field were beginning to perceive anomalies in its underlying paradigm. Many new finite groups were being discovered that could not be handled according to the older techniques. Previously, most groups had fallen into a single category; there had been only one type of group that was different or "strange." Now there were many strange groups. The situation was analogous to that of particle physics where the discovery of new particles has occurred more rapidly than the development of theory to explain them. As a result, the field was going through a period of uncertainty and questioning of older ideas.

Members of this area did not seem to perceive the field as being divided into distinct camps. Over time members of the two large groups of collaborators were less likely to cite the publications of pairs and isolates. One of the large groups cited the other's publications to a lesser extent over time but continued to be cited by that group at about the same rate. However, although there were several distinct topics in the mathematics area (Gorenstein 1964), it did not appear that separate lines of research had evolved as the field expanded. All but one of the ten most frequently cited publications in the area were linked directly or indirectly to one another through citations. Of the 37 next most frequently cited publications, only seven (19 percent) were not linked directly or indirectly either to each other or to the ten most frequently cited publications. Of the 150 publications that cited the ten most frequently cited publications, 40 percent cited more than one of the ten and 19 percent cited more than two.

Russett in his study of the citation patterns of a group of political scientists used factor analysis to identify groups of scholars who utilized the same kinds of materials in their work.[16] He found twelve distinct groups, although with several overlapping memberships. Factor analysis of citations in the mathematics area revealed eleven distinct groups. The percentage of overlapping memberships was much higher than in the political science area. In the latter area, the work of 18 percent of the members was frequently cited by members of three or more groups. In the mathematics area, the

16. The method is described by Russett as follows: "By factoring the matrix where citers are considered as variables and the authors cited as observations, the resulting factors identify groups of scholars who evidence similar choices; people who tend to read and absorb the same kind of materials. Each citer has a loading, or correlation, with each of the major factors. The squared loading indicates the percentage of the variance (variation) in the chooser's pattern of citations that can be explained by the factor. Once we have identified the factors as groupings of choosers, it is then possible to compute factor scores to see what sources are commonly used by the members of a particular group. Thus each man cited has a factor score for each factor, indicating the relative contribution his materials make to the 'typical' citer in the group" (1968: 7–8).

For this part of the analysis of the mathematics group, the members of the sample were scored in terms of the number of times they cited publications by other members of the sample. Citations to publications by sample members not contained in the bibliographies which were used to generate the sample were included in these scores since it did not seem appropriate to score these as citations to 'outsiders.' Citations to junior authors were also scored.

work of 88 percent of the members was utilized by members of three or more groups. Thus, by these criteria, the mathematics area was highly unified intellectually, much more so than the political science research area.

Conclusion

Social factors within a research area affect the dissemination of information within it and the extent to which information is likely to be utilized in later publications. This can be seen by the fact that information that is heavily utilized at an earlier point in time is most likely to be heavily utilized at a later point in time. This can be interpreted either as the effect of the quality of an innovation (S. Cole, 1970) or as the effect of a social influence process that increases the visibility of previously utilized work. A high level of utilization reflects conformity to norms set by the invisible college in the area. Innovations that were more frequently utilized or cited by members of these research areas were those that were utilized or cited by the most productive members of the area. In both areas, work published by relatively unproductive authors in the early periods of their history was more frequently utilized or cited than the work published by such members later on. Relatively unproductive authors who published for the first time in the later stages of the development of these areas had little influence upon these fields.

In both areas, the contagion process that made certain ideas more visible than others decreased as the areas increased in size. In the rural sociology area, ideas appearing during later stages of growth were less frequently utilized than ideas produced early in its history, even though the number of authors publishing in the later periods was larger than in the earlier periods. Groups of collaborators within the rural sociology area became increasingly unreceptive to each other's ideas. Although the sheer numbers of publications appearing in the area may have affected the visibility of new ideas, the ideas themselves may have become more specialized and less original.

The mathematics area remained unified and continued to produce ideas that were frequently utilized, but uncertainty about the directions that research should take may have been a factor in what appeared to be a decline in the amount of innovation as measured by the proportions of publications not citing other publications in the area.

Kuhn (1962) has stressed that scientific change occurs when the perception of anomalies that cannot be explained by a paradigm produces a period of crisis and the eventual substitution of a new paradigm for the old one. This event, which is usually resisted by many scientists in an area, constitutes a scientific revolution. Alternatively, scientific change may frequently occur as a result of the exhaustion of paradigms that gradually become less attractive to outsiders and less able to compete for new recruits. It is possible that many research areas do not move very far beyond the ideas that originally stimulated their growth. Instead, once the implications of these ideas have been exhausted, such fields are abandoned in favor of a new set of seminal hypotheses. Even theoretical physics, one of the most spectacularly successful areas in modern science, has apparently experienced a decline in innovation, according to one knowledgeable participant-observer:

After the thirty fat years in the beginning of the present century, we are now dragging through the lean and infertile years, and looking for better luck in the years to come. In spite of all the efforts of the old-timers like Pauli, Heisenberg, and others, and those of the younger generation like Feynman, Schwinger, Gell-Mann, and others, theoretical physics has made very little progress during the last three decades, as compared with the three previous decades (Gamow 1966: 161).

5
Variations in Scientific Growth

THE GROWTH OF RESEARCH AREAS IN BASIC SCIENCE HAS
been shown to fit the logistic curve. A series of intellectual
and social changes occurs that brings about changes in the
rate of growth. While this pattern is probably the predomi-
nant one in basic science, it may not apply under certain
conditions. According to Kuhn, some fields, particularly
those in the social sciences, lack paradigms. As a result,
members fail to agree about the important problems and the
methods for solving them and knowledge is not cumulative.
Instead of building upon each other's work, they continually
dispute each other's theoretical interpretations and empirical
findings. Controversies that remain unresolved for considerable periods of time may affect the pattern of growth of
knowledge in a research area.

The literature of other types of knowledge, such as humanities and technology, also exhibits patterns of growth unlike that typical of the basic sciences. This suggests that a
different sequence of intellectual and social events occurs in
their development.

Controversy and the Emergence of Schools
Kuhn (1962) has suggested that when two paradigms con-
front the same research problems, the period of crisis is
often lengthy. In fact, many of the supporters of the original

paradigm are never convinced by the new one, although newcomers to the research area may be so convinced. Gradually a new generation takes over the field and the crisis is forgotten. Alternatively, newcomers to an area may not select one paradigm over another and a prolonged period of coexistence may ensue. Whether or not either paradigm produces a period of exponential growth depends upon the extent to which the field is able to attract new members despite the controversy surrounding its paradigms.

Relationships between the followers of opposing paradigms may be characterized either by avoidance or by confrontation. There is some evidence that confrontation was more typical of an earlier period when scientific communities were smaller. Krantz (1969, 1971) has studied examples of both these responses to opposition between paradigms in the discipline of psychology. He describes a brief period of confrontation that occurred in that discipline between 1895 and 1896 when a prominent investigator developed a new paradigm that was incompatible with a previous one associated with an equally important man. The latter reacted with intense hostility. Analysis of the content of four papers that the two men wrote during the period of controversy showed that the basic argument centered around the paradigms themselves, i.e., "how to conduct valid psychological research." Empirical data was not an issue for the most part. Nor could empirical data resolve the controversy because the disputants did not agree about their relevance. "Facts alone could not resolve the controversy since their relevance was determined by the very issues which were at the root of the controversy. In short, facts could only be important when the disputants agreed upon a common ground of discussion" (Krantz 1969: 10).

The controversy that began with "rational" and "scientific" issues degenerated into invective and personal attack designed, according to Krantz, primarily to influence the views of their followers rather than each other. The controversy between the two paradigms was never resolved. Eventually, a new paradigm supplanted both.

Even in that period, however, direct confrontation was rare in psychology. Confrontation can only occur when different theoretical interpretations are applied to the same data. Supporters of different paradigms usually devoted their attention to different research areas. This did not prevent each group from expressing considerable contempt for other viewpoints.

Today, supporters of opposing paradigms in psychology avoid confrontation while studying the same sets of problems by developing separate scientific communities with very few points of contact. In the area of learning, Krantz (1971) shows that the Skinnerians and their opponents constitute two quite distinct scientific communities. The Skinnerians are particularly unreceptive to external influence. Many Skinnerians are former students or collaborators of Skinner or students of his former students. Skinnerian psychologists have their own journal that publishes only material using Skinner's theoretical approach to the study of learning. Skinnerians seldom publish in other psychological journals. This avoids the controversy that would be entailed if supporters of different paradigms attempted to publish in the same journal. Papers by Skinnerians primarily cite psychologists who are identified with Skinner's approach. These findings suggest a group that is closed to external influences and in this sense has some of the characteristics of a "school."

A school is characterized by the uncritical acceptance on the part of disciples of a leader's idea system (Krantz 1971). It rejects external influence and validation of its work.[1] By creating a journal of its own, such a group can bypass the criticism of referees from other areas. The distinction between a school and a solidarity group is somewhat tenuous

---

1. Schools also have similarities to religious sects. The latter break away from the church and build separate organizations, emphasizing aspects of doctrine or policy that they believe have been ignored or misinterpreted by the church. The religious sect is a relatively closed system that resists external influences rather than attempting to adapt to them. Members who deviate from orthodox views on any issue are quickly expelled (see, for example, Coser 1954; Johnson 1964; Yinger 1957).

since all solidarity groups are committed to a particular point of view to which their members are expected to conform. In the phage group of molecular biologists, however, members were continually critical of each other's work (Mullins 1968b and c).

In extreme cases, as described by Back (1971), such groups forgo the attempt to obtain empirical verification for their ideas altogether. Back has described the gradual evolution of the group of researchers devoted to the study of the technique of sensitivity training. From a scientific emphasis upon the evaluation of the technique of sensitivity training in terms of its effects upon participants, both desirable and undesirable, the group has moved toward the acceptance of personal testimonials concerning the effectiveness of the method in place of the collection of quantitative data. Concomitantly, it has attracted increasing numbers of adherents who are neither scientifically trained nor motivated.

It appears that, when two paradigms are applied to the same research area, followers of the two paradigms can avoid confrontation and develop as if they belonged to two different research areas, each of which goes through the stages of logistic growth and contains solidarity groups and invisible colleges. Krantz has documented the rapid expansion of the Skinnerians. The group has grown so large that distinct subgroups have emerged, each devoting their attention to different research interests within the general area. One of these subgroups has recently started a journal of its own. A new journal is being contemplated by another group because the two existing journals are viewed as "too conservative."

Krantz concludes that, in the case of the Skinnerians and their opponents, "there appears to be little desire for either side to shift, assimilate, and integrate the different faiths and the empirical generalizations generated from them." When, on the other hand, prolonged confrontation rather than avoidance is the pattern of interaction between the followers of the two opposing paradigms, one would expect that growth of the field would be inhibited. The generally negative attitude toward controversy held by most scientists would tend to inhibit recruitment.

Kuhn describes instances where confrontation led to the resolution of controversy, that he calls scientific revolutions. In some cases that he describes, quantitative evidence was able to resolve the controversy quickly. In other instances, where the opponents did not agree about the relevance of empirical data, the fact that the controversy was eventually resolved in favor of one of the opposing paradigms may have been due to social factors rather than to the superior qualities of the winning paradigm. Examples of two famous controversies of this sort—the Keynesian revolution in twentieth-century economics and the Darwinian revolution in nineteenth-century biology—suggest what some of these factors are.[2] In both cases, the innovators were middle-aged; both were in their fifties when their innovative work was published. Both were well-known and highly respected prior to the publication of their revolutionary works. This suggests that, when agreement about the relevance of empirical data is lacking, an "elder statesman" of accepted reputation may be required to resolve a serious controversy. Both men had numerous "anticipators." The fact that their ideas were accepted and those of their anticipators were not implies that their own personalities and reputations played a role in winning acceptance for their ideas.

These controversies were not resolved because Darwin and Keynes presented clear and unambiguous statements of their theories. One writer (Hyman 1966: 26) has claimed that: "The evidence to establish the idea of evolution by natural selection inductively was not available in 1859, and many of Darwin's processes turn out on closer examination to be plausible hypotheses and his causes tautologies." Keynes's discussion of his theory is notoriously difficult to understand. It has been said: "Every economist of affirmative intelligence emerges from [his book] with his own view as to its decisive point." Both men presented their theories in polemical terms. Keynes's original statement of his innovative economic theory has been called by an economic historian "a battle cry," and it included a ruthless attack on

2. For descriptions of these revolutions see Darwin (1958) and Lekachman (1964).

earlier economic theories. In contrast, when mathematical evidence is accepted by both parties, major innovations tend to be the work of young and relatively unknown men and to be stated in a nonpolemical fashion.

Logistic Growth and Fashion

Kuhn implies that, in general, knowledge is not cumulative in the social sciences as it is in the natural sciences. He attributes the difference to the absence of paradigms in the social sciences that makes "normal science" impossible. Instead, continual disagreement prevents contemporaries from building upon each other's work. The literature of at least one social science, sociology, however, exhibits rapid exponential growth. In recent years, the literature of sociology has been doubling every three years (see Figure 15), much more rapidly than the literature of physics, which is doubling every eight years (Anthony et al. 1969: 713). This implies that sociologists are influenced by one another in their selection of research problems and are building upon each other's work. If this is the case (and the analysis presented in Chapter 2 suggests that it is), then logistic growth may have different implications depending upon the type of research area.

One implication might be that social factors are much more important in recruitment into research areas in the social sciences than cognitive factors. In other words, such fields attract new members because they become fashionable rather than because of their scientific potential. Fashion is generally viewed by scientists as having a detrimental effect upon the development of science, since it is believed that manpower is recruited from significant research areas for less important ones. Using this orientation, Hagstrom (1965: 177–84) has explained fashion in science as essentially a motivational phenomenon. According to him, scientists who follow fashions select research problems not for theoretical reasons but because they believe that solving these problems will bring them rapid recognition. Unfortunately, the true motivation of an individual scientist is difficult to determine. The desire for recognition, as Hagstrom himself argues, is always an important motive for undertaking scientific work.

Another difficulty with this type of analysis of fashion in science is that it treats it as an individual phenomenon rather than as a social one. It does not provide criteria for evaluating the presence or absence of fashion in any particular area. Scientists themselves disagree about whether or not a particular area is fashionable. The same area that will be described as having considerable theoretical importance by some scientists will be dismissed as merely fashionable by others. It is also possible that scientists who select research areas solely on the basis of social validation by colleagues will not necessarily do so in large numbers, making the criterion of size alone insufficient.

Instead, it is more fruitful to examine fashion as a characteristic of groups rather than of individuals. Fashion represents a case of rapid diffusion of an innovation due to a combination of personal influence among peers and social influence in the form of demonstration effects in which individuals observe the behavior of others with whom they are not personally acquainted. Opinion leaders, individuals who are considered to be especially knowledgeable about the relevant topic, play a role in both types of influence.

As we have seen, exponential growth occurs in fields in which social interaction is present. Thus personal influence is having a role in the rapid diffusion of information in a research area and in attracting newcomers to it. All the same, the inference that newcomers enter the research area solely because they expect to obtain rapid recognition and not because of their perception of the importance of the research problems cannot be drawn. Studies of the diffusion of technological innovations generally show that a combination of personal influence plus evidence concerning the importance or usefulness of an innovation is necessary if widespread acceptance of an innovation is to take place (Rogers 1962: 124–26).

It is likely that, both in the natural and in the social sciences, fashion is a derogatory term used by those who do not accept a new orientation toward a set of research problems. In other words, a period of rapid growth in a research area may be perceived by those outside the area as a fashion. The criteria for evaluating the importance of problems in the early stages of a research area are rarely entirely objective. This is inevitable since thorough assessment of a new approach may take years. As a result, there is bound to be disagreement between insiders who have accepted a new approach and outsiders who consider it trivial and hence "fashionable."

Whether or not scientists in other fields will label a period of exponential growth in a research area as a "fashion" depends partly upon the extent to which very large numbers of scientists enter the field so that the field becomes highly "visible" to outsiders. Exponential growth refers to rate of growth rather than absolute size so that a group could increase in size rapidly without becoming so large that its growth seems remarkable to members of related areas.

If social interaction in the absence of intellectual content were causing the growth of these areas, one would expect that such research areas would be exhausted very rapidly, but this does not appear to be the case. For example, in medical sociology, a specialty that has been called fashionable (Barber 1968), one research area has exhibited a period of exponential growth lasting as long as twenty years.[3] In comparison, in the mathematics area that was studied by the author, the period of exponential growth lasted fifteen years. It seems likely that the growth of all research areas, in the natural and in the social sciences, is "fashion-like" in the sense that rapid growth occurs as a result of social interaction but not true fashion since technical criteria for evalu-

3. Another area grew at an increasing rate for a period of twenty years (exponential growth is growth at a constant rate). These data were computed on the basis of the bibliographies appearing in Freeman et al. (1963, 1971). The labels given to various research areas in the two bibliographies are different in the second edition. For purposes of this analysis, bibliographies 3 and 4 in the first edition and bibliographies 4, 5, and 6 in the second edition were treated as a single research area. Similarly, bibliography 2 in the first edition and bibliographies 1, 2, 7, and 8 in the second edition were treated as a single area. These decisions were made on the basis of the similarity of the content of the items on the various lists. The second of these areas exhibited exponential growth.

ating the information being transmitted through the communications network are available. This makes these processes similar to the diffusion of most technical innovations rather than similar to "pure" fashion in which acceptance is primarily emotional, based upon the gratification of physical or emotional needs and with little element of rationality.

It is possible that differences between natural and social science research areas are due not to the absence of paradigms but to the nature of the paradigms that are available for study. Masterman (1970: 74) suggests that the social sciences have paradigms that are "more trivial and narrow than [each] field as defined by intuition." The operational definitions given by techniques cannot be reconciled with one another so that it is not possible to merge narrow fields into broader ones. For example, this type of situation may exist in the research area that is devoted to the study of small (human) groups. McGrath and Altman (1966) describe a period several decades ago when confrontation between sets of paradigms was frequent in this research area. This period has been followed by one in which a large number of investigators examine narrow topics with little concern for theory or for the implications of their research in relation to that of other researchers in their field:

. . . we see scientists searching for ways to generate research studies more rapidly and efficiently so they can get their product in the public's view. This production speedup is accomplished by . . . development of gimmicks. One form that such gimmicks take is the development of a unique procedure, piece of equipment, or task. A researcher can then employ the procedure, task, or piece of equipment over and over, introducing new variables or slight modifications of old variables, and thereby generate a host of studies rather quickly (1966: 87).

This field has been characterized by logistic growth (see Fig. 11) and by the development of numerous groups of collaborators with apparently little effective communication between these groups (see chap. 3).

Types of Knowledge: Basic Science, Humanities, and Technology

Thorough understanding of the processes of growth in the basic sciences requires some means of distinguishing between this type of growth and the growth of other types of knowledge. There are two types of knowledge that seem to be different from the basic sciences in their patterns of growth. One is represented by the humanities, the traditions of scholarly analysis of the arts, literature, history, and philosophy. The other is represented by technology, scientific knowledge developed specifically for or applied to the solution of practical problems. Unfortunately, information about both of these types of knowledge is limited.

Price (1965b, 1970) has argued that neither the humanities nor technology exhibit the type of cumulative growth that occurs in the basic sciences. He argues that growth of the literature of the humanities is entirely unstructured. New developments are based upon a "random raiding of the entire archive of the literature" rather than upon a small group of recent contributions that is the pattern in the basic sciences. He shows that the percentage of references to recent literature in many humanities journals is very low. This is what would be expected if growth is occurring on the basis of innovations drawn from the entire previous history of the field.

It was suggested earlier that this type of growth in which new developments can be based upon any previous development implies the least amount of social organization. If this is the case, growth of the literature in these areas should be linear rather than exponential since social influence is having a minimal effect upon the diffusion of ideas. Analysis of the growth of publications in English literature from 1923 to 1967 reveals a linear pattern of growth until 1939, followed by a very slow rate of exponential growth (doubling every seventeen years rather than every ten years as in the basic science literature (see Fig. 16).

On the basis of these findings, it seems reasonable to speculate that the types of puzzle-solving paradigms that occur in the basic sciences are generally absent in the humanities. Orientation paradigms, or expectations concerning the nature of phenomena, may be present. For example, in recent decades one type of orientation has influenced one of the humanities, English literary criticism. Crews (1970) has analyzed the tendency among literary scholars to reject systematically all nonliterary explanations of literary works. According to Crews: "Each critic is free to adopt the 'approach' that suits his fancy, and most of the approaches prove to be little more than analogical vocabularies lending an air of exactitude to whatever the critic feels like asserting." He suggests that the field does not function as a scholarly discipline in which different approaches to the understanding of a particular phenomenon are compared and evaluated. Thus it is not surprising that literary scholars do not cite each other's work. They are not concerned with creating a common body of knowledge of their subject. This "approach" stems from the work of leading literary critics and is embodied in at least one book that specifically defines norms for literary criticism. Crews implies that this type of orientation is not the only possible approach to the subject. Should this orientation be replaced by one that stressed the development of common theoretical frameworks, one would expect to find the development of groups of scholars using different types of theories to explain literary phenomena. If this occurred, in time references in the literature of the area would presumably exhibit greater structure and a higher proportion of references to recent items.

The less precise the paradigm, the more room for disagreement, and the less likely that large numbers of researchers will accept it and choose to work on it. Thus one would expect that social circles focusing upon particular research areas in the humanities would be smaller than those in the natural sciences. Recent data reviewed by Roberts (1970: 298) provides some evidence for this hypothesis. A study of researchers in the language sciences found a "surprising" number of research areas in which less than a dozen workers were engaged.

Technology appears to be more heterogeneous than the humanities in terms of the variables being considered here.

The term is applied to a very large group of activities, some of which resemble basic sciences in that primarily understanding of phenomena is being sought, while others are oriented toward the development of specific techniques or products. Studies of citations indicate considerable variation in the extent to which technical publications utilize basic science publications (Marquis and Allen 1966). Some branches of technology appear to be independent of basic science, while others draw on its findings.

The problem is further complicated by the fact that it is notoriously difficult to differentiate satisfactorily between basic and applied science. A recent study (Illinois Institute of Technology Research Institute 1968) distinguished two types of nonbasic science: (1) mission-oriented research—"performed to develop information for a specific application concept prior to development of a prototype product or engineering design"; and (2) development—"involving prototype development and engineering design directed toward the demonstration of a specific process or product."[4]

The growth of technology is difficult to study because the literature is not a complete record of the innovations that are produced. The technologist is less motivated to report his innovations in the literature because he is afraid that others will exploit their commercial possibilities before he is able to do so. For this reason, the recent study by the I.I.T. Research Institute is especially useful since, in tracing the history of the development of five major technological innovations, it utilizes research and development "events" (i.e., discoveries of information in some way related to the development of such an innovation) rather than publications. Although it is not clear to what extent one can generalize from the history of the development of these five major innovations to the development of all technological innovations, the findings of this study suggest that the growth of technological areas is different from that of basic science areas. A major technological innovation is based upon a

4. Leach (1971) argues that basic and applied research can be distinguished only on the basis of the type of organization that is performing it; for example, university, industry, or government.

large number of findings resulting from all three types of scientific activity, basic, mission-oriented, and development. While basic research generally precedes the latter two types of research, it is sometimes stimulated by technological innovation. Much of the diffusion of information from one type of science to another is by personal communication rather than through the literature.

The I.I.T. study suggests that the technologist is primarily interested in finding puzzle-solving devices that will permit him to reach a specific goal. Thus he is not likely to be interested in "orientations" that define a range of problems and provide an over-all perspective. The same study also indicates that the growth curve of the development of a major technological innovation is different from that of a basic science research area. There is a relatively slow accumulation of information over a period of several decades followed by a rapid acceleration of activity just prior to the development of the major innovation. The difference between the two types of growth pattern is underlined by the fact that in basic science the goals of the research area are defined early in its development and stimulate its development. In applied science, the possibility that an important innovation can be developed is often not recognized until a few years prior to its invention.

Conclusion

Different types of knowledge exhibit different patterns of growth. Logistic growth is characteristic of basic science but not of humanities and technology. It occurs both in the natural and in the social sciences. In the social sciences it is not simply the result of a process of social influence or fashion but apparently reflects the acceptance of innovations based upon judgments concerning their usefulness or validity to the adopter since periods of exponential growth continue for substantial periods of time. Tendencies toward a high level of fragmentation of interest in some social science research areas may reflect differences in the nature of the paradigms in the social sciences compared to the natural sciences but do not imply the absence of paradigms.

In basic science, controversy does not necessarily inhibit growth. Opposing groups committed to different approaches split off and develop separate bodies of work. The rate of growth of knowledge *is* affected by the absence of social interaction and absence of agreement about approaches within the group studying a set of problems. This type of situation appears to be characteristic of the humanities and sometimes occurs in the basic sciences (see chap. 2).

Technology is characterized by short spurts of rapid growth generated by the realization among groups of researchers that a particular innovation can be developed. In the process, they often draw upon knowledge from both basic and applied science. In technology as well as in basic science, social interaction facilitates the diffusion of information, but little is known about the nature of this type of social organization.

6
Interactions between Scientific Communities

IN PREVIOUS CHAPTERS, SOME RELATIONSHIPS BETWEEN communities of scientists and the growth of scientific knowledge have been documented. It has been shown that each scientific community concentrates its attention upon a particular set of problems. A few scientists are attracted to these problems by an interesting piece of work, and they in turn convince others to join them through recruitment and training, as collaborators, or by the indirect influence of their publications. A social community with a distinctive structure appears and for a time expands rapidly while producing a considerable volume of work.

Until now, the factors that influence the appearance of these paradigms have not been considered. Are scientific communities relatively closed to external influence, as Kuhn and Kroeber imply? Do autonomous communities develop separate streams of thought with only a minimal exchange of ideas? This chapter will explore the nature of the cognitive and social connections between research areas. Evidence will be presented that suggests that social and cognitive influences flow across research areas at all stages of their growth. It will be argued that this openness to external influence plays an essential role in the process of innovation in scientific communities.

Scientific Communities: Open or Closed?

If it is true that acceptance of a paradigm has the effect of narrowing a scientist's perspective so that he concentrates his attention solely upon the problems it raises, this would imply that once a field has developed its first paradigm, further change occurs entirely within that framework. Exchange of ideas between research areas would be inhibited by the nature of the paradigm itself. Only those who view a subject in the same way are able to exchange ideas about it.

If research areas were completely closed to external influences, scientists would be divided into small groups, sharing the same interests, speaking only to each other, and reading and citing only each other's work. If this were the case, science would consist of hundreds of disparate groups, none of which would have any communication with or relevance for each other.

If members of research areas are aware of and have contacts with scientists working in other research areas, this suggests that research areas are not closed communities, unreceptive to external influences. Respondents in the rural sociology and mathematics areas had numerous contacts with scientists who had not published in these areas. In the rural sociology area, respondents were as likely to choose a scientist who had not published in the area as they were to choose a scientist who had published in the area. Out of a total of 1,351 choices made by all respondents on all the different types of sociometric ties examined,[1] outsiders represented 51 percent and members of the research area 49 percent. Crawford's findings (1970: 32) for sleep and dream research were somewhat similar: 58 percent within the area and 42 percent outside the area. In the mathematics area, 436 choices were made, 28 percent for outsiders and 72 percent for members of the area.[2]

1. Published collaborations were excluded from this analysis since by definition members of the area could not have had published collaborations in the area with outsiders.

2. The total number of choices is substantially smaller in the mathematics area compared to the rural sociology area because the group itself was smaller (102 members compared to 221) and fewer types of ties were examined (see chap. 3).

At first glance, these findings seem to contradict previous conclusions about the existence of social organization in the area. It was clear, however, that relationships with outsiders were different in character from relationships between members of the area. In the rural sociology area, the majority of "outsiders" were selected only once.[3] Eighty-four percent were chosen no more than twice. Three percent were named more than five times. On the other hand, 12 percent of the members of the area were named more than five times. Seven percent of the members were named more than ten times. Since only one outsider was named more than ten times, there did not appear to be a group of outsiders whose influence had similar weight. In the mathematics area, the majority of "outsiders" were mentioned once;[4] none was mentioned more than four times. On the other hand, 17 percent of the members of the area were named more than five times and 6 percent more than ten times, figures that are very similar to those for the rural sociology area.

Sociometric choices of members of these areas seldom converged upon individuals outside the area. Instead, the areas had internal bonds that gave them structure and identity and numerous links to other areas in the form of single ties to a large number of individuals. The latter provide obvious channels for the diffusion of information both from and to the area.

Studies of the extent to which scientists writing on a particular topic cite references related to different topics show a similar pattern. Depending upon the subject, approximately half the references in a sample of scientific articles are to relatively recent papers in the same field. The remaining references are to papers from many different scientific fields.

Unfortunately, most of the available data deal with very broad subject categories rather than with research areas, although they give some indication of the existing patterns. Studies of "subject dispersion," defined as the degree to which the literature on a particular subject cites publications from different subjects (Stevens 1953), suggest that all re-

3. See Table 14.
4. See Table 15.

search areas rely to some extent upon other fields. Social sciences have a higher level of subject dispersion than do the natural sciences, but the latter rely more heavily upon closely related fields than do the social sciences.

Brown's data (1956) give a broader picture of the connections between disciplines. Using journals published in 1954 in eight disciplines,[5] he examined the subject areas of the journals that these journals cited. His data provide some indication of the structure of interdisciplinary relationships. Mathematics, for example, was a relatively "isolated" discipline. None of the source journals in the other seven disciplines that Brown examined cited mathematics journals to any significant extent.[6] The mathematics journals themselves exhibited weak ties to physics and applied physical sciences. A citation study by another author (Meadows 1967) suggests that astronomy is similar to mathematics in these respects.

The remaining disciplines fell into three distinct subgroups, the physical sciences, the biological sciences, and their related applied disciplines, such as engineering, medicine, and agriculture. For example, source journals in physics had weak ties with several interdisciplinary areas associated with physics such as astrophysics and geophysics and a strong tie with chemical physics. Geology depended to a considerable extent upon the literature of chemistry and physics. The biological sciences (physiology, botany, zoology, and entomology) also made considerable use of each

5. The disciplines were mathematics, physics, chemistry, geology, physiology, botany, zoology, and entomology.

6. For purposes of this analysis, a strong tie between disciplines was indicated if a source journal cited any journal in another discipline more than twenty-five times. A weak tie was indicated if a source journal cited a journal in another discipline between ten and twenty-five times. No attempt was made to characterize interdisciplinary journals such as *Nature* and *Science* for purposes of this analysis. Earle and Vickery's data (1969) is in general similar to this data with the exception of mathematics, which they find to be heavily cited by other disciplines. These other disciplines are technology and the social sciences, not the disciplines being considered here.

other's journals. They also cited and were cited by agricultural and medical journals. Chemistry cited and was cited by journals in both the physical and the biological sciences. Brown reports that the chemical literature is also extensively used by engineering and medical and agricultural sciences.

Studies of comparable scope do not exist for the social sciences, but judging from citations contained in the *American Sociological Review* in 1965, sociology has weak ties with a number of other disciplines (Broadus 1967). Social history, medicine, and economics accounted for the highest proportion of the citations to other disciplines, with statistics and psychology following closely. Anthropology and the natural sciences played the smallest roles in the sociological literature, being exceeded slightly by two of the humanities, philosophy and education. Another study of citations in psychological and sociological journals (Parker et al. 1967) found fairly strong connections between social psychological journals and sociological journals, suggesting that some diffusion of ideas occurs between that area of psychology and sociology. A citation study of the literature of geography showed that it has its closest interdisciplinary ties with the literatures of history, economics, and geology (Stoddart 1967).

Although these studies discern only the interrelationships between disciplines rather than research areas, each field appears to be related to a few others, but in such a way that all fields are interlocked in a kind of honeycomb structure. There are, however, variations between disciplines, with mathematics being almost entirely self-contained, the physical sciences relatively so, and the biological sciences least of all. Chemistry appears to act as a bridge between these latter two sets of disciplines with information moving in both directions.

The differences in the extent to which disciplines draw upon the work of other disciplines may be related in part to how narrowly a question can be posed. In some disciplines, it is more difficult to disentangle a single aspect of the subject matter from the remainder. This seems to be particularly true in the biological and social sciences and less true in the physical sciences and mathematics.

Polanyi (1962: 59) seems to have had a sort of honeycomb structure in mind when he describes science as consisting of chains of overlapping neighborhoods extending over the entire range of science. Each scientist understands the paradigm of his own field and enough about that of neighboring fields to be able to evaluate the research produced in those fields, but that is the extent of his understanding.

Another sociometric study appears to corroborate Polanyi's suggestion. Using the handbook of scientific biographies, *American Men of Science,* Mullins (1968a) drew a random sample of fifty biological scientists, each of whom had a Ph.D. and was affiliated with a university. The fifty scientists were members of seven biological disciplines. Questionnaires were sent to these men asking them to name scientists "working in the same or closely related areas and with whom they had contact about research within the last year." Using a snowball design, scientists named by this original group were also sent questionnaires. The process was repeated twice. While Mullins found numerous small clusters of scientists who were linked by direct choices, some of which formed larger groups linked by indirect choices, he also found that his networks with starting points in seven disciplines reached scientists in sixty-four different disciplines, ranging from anthropology to geology, from virology to chemistry.

Types of Cross-Fertilization

Kuhn may have been led to his view of the closed nature of scientific communities by his emphasis upon revolution as a mode of change, which implies transformation within a group rather than a "takeover" from outside. If one views research areas as being exhausted and replaced by new topics, the emphasis shifts to exogenous sources of change. A process of cross-fertilization occurs in which ideas from one field generate a period of rapid growth when applied to another field. Holton (1962) compares science to a tree whose limbs are constantly branching to create new fields and subfields of knowledge. He suggests that new fields are created in part by the discovery of linkage between old fields. A paradigm that has been developed in one field is applied to a different field. Another type of growth occurs when ideas are drawn from a variety of fields. According to Holton (1962: 383): "Inputs for a lively research topic are not restricted to a narrow set of specialties but can come from the most varied directions." Juxtaposition of ideas from several different fields may produce a new paradigm that is applied to a newly defined research topic.

March (1965), who used citations to investigate the intellectual origins of organizational theory, discovered such a pattern of innovation. Using a set of twelve representative volumes from the current literature on organizational theory, he found that they frequently cited thirty-three earlier works, most of which had appeared in the previous two decades. There was no significant overlap in the citations made by these "classics"; evidently, the books that had stimulated the growth of that area originated in a variety of fields.

Why is cross-fertilization between research areas possible and how does it occur? Holton argues that exchange of ideas among rapidly proliferating fields is possible because they share certain concepts.

What makes certain concepts important is their recurrence in a great many successful descriptions and laws, often in areas very far removed from the context of their initial formulation. An idea has to stand the test of wide application. It is hard to tell when it is first developed what will happen to it (1952: 233).

This implies the existence of a common language in science that is similar to the metaphysical or orientation aspects of a paradigm. These concepts can be operationalized, quantified, and are applicable to diverse fields of inquiry. Mullins (1968a) found some evidence of the existence of this common language of concepts. The scientists with whom his respondents exchanged information were likely to share with one another a similar "orientation" toward research materials. For example, research can emphasize chemical or physical aspects of nature, structure or process, substance or technique. It seems likely that when this common language is lacking, there is not sufficient mutual understanding for borrowings to occur.

If scientists are able to locate ideas in other fields, there must be conditions that facilitate the transmission of ideas from one area to another. Several of these have been identified and will be discussed here: (1) information-seeking— the strategies that scientists use in seeking information for their research; (2) reference scattering—the distribution of articles on different subjects in journals belonging to various specialties; (3) specialty hybrids—scientists who work on more than one topic and as a result develop ties with scientists in different areas; (4) interdisciplinary research areas—topics that are studied by scientists from several disciplines; and (5) role hybrids—scientists who have positions that bring them into contact with both basic and applied research problems.

Back (1962) implies that the strategies that scientists use in seeking information foster the diffusion of ideas. He suggests that scientists engage in two types of information-seeking. The first method consists of a highly directed search for particular types of information. The second method is described as undirected searching in a "wide but bounded" field. Back argues that this increases the breadth of the scientist's knowledge and is therefore stimulating to his work. The discrepant ideas that he meets are stimulating and prevent him from becoming too narrowly specialized. A study of the reading habits of chemists and physicists revealed that they devote between a quarter and a third of their reading time to journals in other disciplines. Sixty-five percent of the chemists and 40 percent of the physicists said that their reason for reading scientific journals was "undirected browsing" (Anthony et al. 1969: 734, 752).

The findings from the numerous descriptive studies of information-seeking by scientists (Menzel 1962, 1967) can be summarized most economically in terms of these two modes of behavior that appear to be related to the structure of scientific knowledge. A scientist explores his own field through directed searching. The many links to other fields are the results of undirected searching that has revealed unanticipated foci of relevance beyond the core topic.

To some extent, the literature itself may stimulate this process by means of the location of journal articles. Studies of "reference scattering" reveal that about half the articles on any subject are concentrated in a dozen or so journals; the rest are widely scattered among hundreds of journals on other subjects (Swanson 1966). For example, "mainstream" geography has been described as consisting of a relatively small number of important journals, which are surrounded by a diffuse and peripheral literature extending into many fields of knowledge (Stoddart 1967). A scientist who sees a publication from another area in a journal in his own field may thus find his way into the literature of a new field.

This undirected searching for information might be less unpredictable if the social ties that link areas of knowledge were better understood. In some disciplines, many scientists work in more than one research area. It is not unlikely that cross-fertilization of ideas occurs by this route. The small group of productive scientists that provides the necessary cohesion within a research area may have ties with the most productive scientists in other areas, thus providing channels for the diffusion of ideas. Some support for this hypothesis is provided by the fact that, in the study of the rural sociology area, the most productive scientists had more ties with scientists who had not published in the area than did the less productive scientists. Seven of the twelve outsiders who were chosen more than five times by members of that area had served as president of the American Sociological Association or the Rural Sociological Society. These scientists were well known in the discipline as a whole and undoubtedly highly productive.

A study of the "visibility" of a group of physicists to their colleagues in that discipline provides further evidence for this hypothesis (Cole and Cole 1968). The visibility of a physicist was defined in terms of the extent to which other physicists were familiar with his work. They found that physicists whose work had been frequently cited were known to almost all physicists regardless of their specialty. In other words, frequently cited physicists were only slightly better known to physicists within their own specialties than to physicists in other specialties.

Groups of scientists from different disciplines who are studying the same empirical phenomenon may also be influenced by each other's research. In time, if their interests converge sufficiently, they may develop a common paradigm. An economist has described an example of such a situation:

It becomes far easier, and more interesting as well as more productive, for the economist who works with non-market decisions to communicate with the positive political scientist, the game theorist, or the organizational theory psychologist than it is for him to communicate with the growth-model macroeconomist with whom he scarcely finds any common ground (Buchanan 1966: 181).

If scientific fields are like "overlapping neighborhoods" or like "fish scales" (Campbell 1969), there must be many interdisciplinary research areas of this sort. Campbell is concerned that disciplines impose artificial boundaries between research areas. Social pressures such as the availability of professional positions and the desire for recognition tend to keep scientists from entering those peripheral areas of disciplines. These areas tend to be underpopulated because the scientist is more likely to be rewarded for doing research that is central to his discipline.

Ben-David (1964), relying primarily upon historical examples, is also pessimistic about the ease with which ideas are transmitted across disciplinary boundaries and from basic to applied science. If interdisciplinary specialization is required in order to develop an innovation, the organization of university departments often inhibits it. It is usually necessary for the graduate student to be identified with a single department. In cases where joint programs have been developed, the new hybrid often has difficulty finding a job and usually has to take a job in a department representing one rather than both his disciplines. Since research facilities are controlled by departments, it is difficult for young men to obtain their use in order to take the first steps in a borderline field.

If the innovation leads to the development of a new specialty or discipline, then the existing academic system must be capable of expansion, of creating new positions and new departments in order to absorb the innovation and permit it to develop. Ben-David presents a number of historical examples of resistance to such innovations in university systems where new positions could not be created easily.

The diffusion of information across two existing disciplines is different from the problem of institutionalizing a new discipline in the face of competition from older ones. The former is more common in contemporary American science.[7] While interdisciplinary specialization may remain problematic, whether or not ideas are diffused from one discipline to another seems to depend in part upon factors affecting the "visibility" of research in one discipline to members of another discipline. For example, Dahling (1962) traced the appearance of references to Shannon's information theory, produced in the field of communications engineering, in the literatures of other disciplines. Seven years after the initial publication describing the innovation, references to it had appeared in publications belonging to nine disciplines: psychology, physiology, optics, physics, linguistics, biology, sociology, statistics, and journalism. Dahling argues that the idea was diffused rapidly because its relevance was immediately grasped. The problem of measuring information transmission was one that had attracted the attention of many people in many different fields, and when a solution appeared it was quickly recognized. Shannon's eminence and the prestige of the journal in which the idea was published (the *Bell System Technical Journal*) contributed to the speed of adoption. The idea was quickly utilized in its

7. There is some evidence that the organization of European science has changed less in the last few decades than that of American science (Crane 1971).

parent field and spread rapidly to other fields that shared an interest in communication. Dahling also stresses the importance of "centers," both academic and nonacademic, that acted as focal points, bringing people from different disciplines together and exposing them to these ideas.

In another interdisciplinary area, that of the diffusion of innovations, researchers in different research areas were unaware of each other's activities in the early stages of the development of these fields (Rogers 1962).[8] In one instance, two groups of researchers in different disciplines were unaware of each other's existence although they were located in the same university. Later when studies grew in importance and quantity, groups working in different disciplines began to cite each other's work more frequently (Rogers and Stanfield 1966). Buchanan (1966) implies that ideas are diffused among the various disciplines that are concerned with "nonmarket decisions." March (1965: x–xii) provides evidence of cross-disciplinary diffusion in organizational research, a field that includes participants from at least six disciplines. This type of exchange of ideas, however, does not always occur in interdisciplinary areas. A study of the relationships between geography and other disciplines (Mikesell 1969) indicates that some cross-disciplinary diffusion occurred in five cases out of eight, but the diffusion was often unidirectional with geography on the receiving end. The small size of geography as a discipline meant that its achievements were not rapidly recognized by members of other disciplines. Members of related disciplines were likely to be most aware of theories that geographers themselves considered outdated. In spite of the fact that they faced a wide range of common problems, geographers and members of related disciplines generally failed to develop a sustained commitment to common substantive issues.

Another study (Davis 1970) found lags of ten to twenty

8. The diffusion of innovations has been studied in several academic disciplines (anthropology, economics, geography, psychology), in two research specialties within sociology (medical sociology and rural sociology), and in several applied fields (journalism, communication, consumer behavior, and industrial engineering) (Rogers 1962; Rogers and Stanfield 1966).

years between the development of important advances in fields related to psychiatry and the appearance of references to such work in the literature of psychiatry. The author provides no explanation for this phenomenon, but it is possible that psychiatrists, as clinical specialists, are somewhat isolated from the basic scientists who conduct research on related topics.

When ideas from one area are being applied to another area that has not been active previously, little resistance to the innovation is likely to occur. There will not be a group of scientists already committed to a previous approach. Mulkay (1969) argues that much innovation in modern science is of this type and consequently the amount of resistance to innovation is low. But when new ideas are developed in an already active area, resistance is much more likely. In this instance, the response on the part of the innovators is often to isolate themselves from those who disagree. Hagstrom (1965) has described the processes of segmentation and differentiation that occur when a new and deviant specialty is being developed.

The diffusion of ideas between basic science and applied science and technology is also identified as a problem by Ben-David (1960, 1964). Again citing historical examples, he points to the resistance by academic scientists to the development of the fields of bacteriology and psychoanalysis, new disciplines that emerged outside the university. He suggests that innovations that involve a fundamentally new view of phenomena often come from academically marginal situations and from individuals who are "role hybrids" in that they participate both in basic research and in its application rather than from academic scientists. This is because some important discoveries are suggested by practical rather than theoretical considerations.

Nonetheless, a study of the development of five recent major technological innovations (I.I.T. Research Institute 1968) presented documentary evidence of considerable exchange of ideas between the basic sciences, the applied sciences, and technology. This influence moved in both directions. Applied science and technology drew upon a wide range of basic sciences in the development of these innovations, and some research areas in basic science developed and grew very rapidly as a result of innovations in related applied science and technological fields.[9] The report stresses the role of personal influence in the diffusion of ideas from one type of scientific activity to another. In many instances of this type of intellectual exchange personal communication played a role. This is not surprising since the applied scientists and technologists were drawing upon such diverse sources. Personal communication would be expected to be an important factor in locating relevant materials in unfamiliar areas. This suggests the existence of individuals in each type of scientific activity who are receptive to ideas from the other type of activity.

The preceding discussion has dealt with the role of innovations from other research areas in the origins of new research areas. It could be argued that once an innovation from another area has been assimilated, a research area becomes closed to further external influence. Back (1962) suggested that when scientists in an area abandon nondirective searching for new ideas, the level of innovation in the area declines. To what extent are research areas open to ideas from other areas throughout their development?

In the rural sociology area, the number of scientists mentioning *only* outsiders as influences upon their selection of research problems in the area decreased from 38 percent among scientists entering before 1951 to 9 percent among scientists entering after 1956.[10] All the same, examination of the *total* number of choices by members of the group with respect to influences upon the selection of problems in various periods revealed that, among scientists entering the area

9. Other examples of this phenomenon appear in Marquis and Allen (1966). Brown's data (1956) discussed above and Earle and Vickery's study (1969) of citations in the literature of science and technology also show considerable interdependence between the two categories. The latter study found that science sources cited technology sources 29 percent of the time, while technology sources cited science sources 14 percent of the time. For engineering alone, such sources cited science sources 22 percent of the time.

10. The date of a scientist's first publication in the area was used as the date of his entry into the area.

prior to 1951, 42 percent of the choices were for members. Between 1951 and 1955, the figure was 38 percent. Among scientists entering between 1956 and 1960, 57 percent of the choices were for members. Among scientists entering after 1960, the comparable figure was 48 percent. During the height of the field's growth (1956–60), the amount of external influence apparently declined to its lowest point, but the proportion of external influence (43 percent) was still substantial.

The pattern was somewhat different in the mathematics area. Here the proportion of external influence steadily declined, although it never disappeared entirely. In response to the question seeking influences upon their selection of research problems, 27 percent of the choices made by respondents entering before 1950 were for members of the area. Among respondents entering the field between 1950 and 1959, 61 percent of the choices were for members. Among those entering after that period, 79 percent of the choices were for members. These findings are consistent with subject dispersion in citation references in this discipline.

Conclusion

Mapping the cognitive and social relationships between research areas will be the next major task for the sociology of science in order that the circulation of ideas across research areas can be understood. Research areas seem to have tendencies toward both a high degree of specialization and toward receptivity to external ideas.

Swanson claims that a significant proportion of the scientific literature on any topic appears in such a large number of publications that it is almost impossible for any individual to locate all of it. Fortunately, the remainder is located in a few "core" journals. If all the literature on a particular subject were so scattered and if the scientists themselves were not in communication with one another, it would be almost impossible for scientists to build on each other's work. The existence of a "core" of journals in the literature and of scientists in the research area provides a kind of repetition in scientific communication insuring that certain ideas will be repeated and emphasized sufficiently so that the scientists who are interested in these problems will be sure of receiving at least some of the currently important messages and therefore continue to do research on these problems.

The exchange of ideas between members of different research areas is important in generating new lines of inquiry and in producing some integration of the findings from diverse areas. Some degree of closure is necessary in order to permit scientific knowledge to become cumulative and grow, while their ability to assimilate knowledge from other research areas prevents the activities of scientific communities from becoming completely subjective and dogmatic.

7

The Structure of Science: Implications for Scientific Communication

CONCERN WITH THE SPEED AND EFFICIENCY OF SCIENTIFIC communication has produced a number of innovations designed to accomplish these goals. Some involve reorganizing the communication patterns within research areas (Dray 1966; Libbey and Zaltman 1967). At least one involves a radical change in the communication structure of an entire discipline (Van Cott 1970). In this chapter, such innovations will be evaluated in terms of the model of scientific growth that has been presented here.

Information-seeking and the Formal Communication System The informal communication system in basic science in which knowledge is disseminated through personal contacts has been described in an earlier chapter of this book.[1] A formal or public science communication system also disseminates information after having evaluated and validated it for the scientific community. The first step in this process of evaluation usually occurs when preliminary results of a piece of research are presented at meetings of professional associations. The evaluation process is generally least rigor-

1. Price (1970) has stressed that basic science, technology, and nonscience are all different social systems. Each system has its own machinery for evaluation and dissemination of information, and each one must be analyzed separately from the others.

ous here. A somewhat more intensive review is undertaken by the editors of journals who publish papers on the basis of recommendations by referees. In the social sciences, the formal communication system also includes books. The latter are less frequently used in the natural sciences, probably because of the rapidity with which knowledge becomes obsolescent in those areas.[2] Review articles that constitute a special type of journal article represent a more thorough winnowing process in which papers in an area are weighed and evaluated in comparison with one another. Knowledge that enters textbooks probably has survived all of the previous hurdles in order to attain scientific respectability.

This structure has gradually evolved over the past several hundred years. Books and journals were the first to emerge. Meetings of professional associations and review articles appeared within the last century. More recently, another type of structure, concerned not with the evaluation of knowledge but solely with its dissemination, has been developed. This consists of abstracting and indexing services and includes the *Science Citation Index*. The latter permits a scientist to locate all the articles in any year that have cited an article that is of interest to him. In this manner, starting with one article, he can trace subsequent research on the same topic.

One of the difficulties with altering the system in which scientific information is communicated is that each aspect of it has a different and essential function. The informal communication system is concerned with disseminating new information. The formal communication system of journals, books, and review articles first evaluates knowledge and, second, disseminates it. Finally, the abstracting and indexing services are concerned solely with disseminating information and do not attempt to evaluate it.

Admittedly, the steadily expanding volume of scientific information makes it increasingly difficult for the scientist to

2. For example, a study of citations appearing in the literature on plasma physics (East and Weyman 1969: 163) found that 13 percent of the citations referred to books. Broadus' study (1967: 19) of citations appearing in the *American Sociological Review* found that approximately 61 percent referred to books.

locate the information that he needs for his research. This problem needs to be evaluated in the context of the structure of scientific knowledge. Since science is organized in such a way that the scientist can utilize information both from within his own research area and from many others as well, he is unlikely ever to have access to all the information that potentially he could use. The problems that he faces in obtaining the information he needs vary depending upon whether he is seeking it within his own research area or whether he needs information from another research area or even from another discipline. In general, he is likely to have more difficulties when he seeks information outside his research area than he does when he seeks it within his own field.

Within his own research area, the scientist's awareness of existing research, published or unpublished, depends upon his own position in its social organization, the size of the area, and the amount of agreement among researchers in the area about the labels they are using in defining their research. These factors also affect how rapidly he obtains information. Speed is important since unpublished research, when published, could preempt his own discoveries.

Scientists who have no informal contacts with the large groups of collaborators in their research areas have the greatest difficulty in locating information since, as we have seen, the latter serve to link the smaller groups and isolates. If sizable groups of collaborators represent only a minority of the scientists in an area, they probably cannot perform this function very effectively. The larger the area, the less likely that all members will be aware of each other's published and unpublished research. "Visibility" is also diminished during the period when an area is undergoing rapid expansion and many new scientists are entering the field within a relatively short period of time.

The scientist's search for information is also affected by the extent of agreement among researchers in his area regarding the definition of a field, the "label" that they assign to it for purposes of identification. For example, in the two research areas I studied, scientists whose papers were included in comprehensive bibliographies compiled by leading scientists in the area showed different levels of agreement regarding the definition of the areas. Two-thirds of the respondents in the rural sociology area classified their publications under the same category (diffusion of agricultural innovations), while over 92 percent of the respondents in the mathematics area stated that their papers were concerned with some aspects of group theory or finite groups.[3] The greater the agreement about labeling in an area, the more frequently authors will utilize similar descriptors as titles, thus permitting prompt recognition by other authors of the relevance of their subject matter.

The farther the scientist moves from his own area of specialization, in seeking information, the more he must rely upon the formal communication system of meetings, journals, books, and abstracting and indexing services. Some of his difficulties in dealing with the formal communication system stem from the fact that research areas are in a constant state of flux. The labels that scientists assign to them do not necessarily correspond to the categories used by journals or indexing and abstracting services (Slater and Keenan 1967–68). The scientist who is seeking information outside his own research area is particularly unlikely to be aware of the current terms being used in less familiar areas and the categories under which this material is being classified in indexing services. He may overlook material of potential relevance because he does not expect to find it classified under certain categories. This problem is increased when knowledge is becoming obsolescent very quickly, because this usually implies that the labels for research topics are also changing rapidly. For example, Brown (1956: 22, 83) observes that physics became almost a new science between 1930 and 1950. He found that the proportion of citations in journals to volumes published between 1944 and 1953 was highest for physics (75.6), intermediate for physiology and

3. Respondents were presented with citations to their own publications in these areas and asked to name the research specialty and problem area to which these papers belonged. The sources of the citations were not identified.

chemistry (61.6 and 58.1), and lowest for mathematics (47.6) and the biological sciences (33.9 for zoology).

The difficulties associated with locating materials within one's own discipline also depend in part upon the compactness of the literature in the discipline. Brown (1956) defines this in two ways: the number of journals to which citations in journals in the discipline referred and the number of distinct subfields within a discipline that have their own journals. Presumably it is easier to locate information in disciplines in which most of the literature is contained in a small number of journals that belong to that discipline than it is in disciplines in which the literature is scattered across many journals, some within and some outside the original discipline. Information may also be found more readily if distinct subfields do not exist in the sense that, when subfields do exist, members of the discipline may be inclined to define information in other subfields as not relevant when in fact it is.

By these criteria, mathematics is a relatively compact science, and the biological sciences are not. Brown found that journals in mathematics referred to 179 journals while journals in zoology referred to 663. Physics and chemistry were relatively compact sciences and were also characterized by a large number of citations to a single journal (*Physical Review* and the *Journal of the American Chemical Society,* respectively). Similarly, mathematics has relatively few journals representing distinct subfields while the biological sciences and geology have many.

These findings suggest that the location of information within a particular discipline is easiest in the field of mathematics, which is both compact and stable. The situation is more difficult in physics and chemistry and in the biological sciences but for different reasons. Physics and chemistry are fairly compact but not stable, while the biological sciences are relatively stable but not at all compact. For example, Brown describes zoology as

. . . a complex science. It includes many subfields, some of which in so far as the literature is concerned may be considered as independent sciences with their own abstracting and indexing journals. The citations in zoological source journals are to a considerable number of serials . . . . The literature of zoology and its subfields is so extensive that it is difficult for any one library to make inclusive collections (1956: 132).

Other data (Anthony et. al. 1969: 744) suggest that physicists and chemists prefer different methods in obtaining information for their research. Physicists placed a higher ranking than did chemists upon informal sources of locating information such as conversation and correspondence. This may possibly be explained by a somewhat greater need among chemists for information from specialties other than their own, as reflected in a high level of "undirected browsing" (see chap. 6).

These problems are magnified when the scientist seeks information outside his own discipline. First, the scientist is not only likely to be unaware of the categories that researchers are currently using in research areas in other disciplines, but the categories used by abstracting and indexing services also may be unfamiliar to him. Since abstracting and indexing services tend to index material from the viewpoint of a single discipline, a member of another discipline is often unable to obtain information from them (Baker 1970: 742). Second, he is unlikely to be a member of the informal communication system in research areas in other disciplines. One exception to the latter may occur in interstitial areas where several disciplines examine the same phenomenon from different points of view. Under these circumstances, the scientist is more likely to establish personal contacts with scientists in other disciplines.

Some of the conditions that facilitate diffusion of information across disciplines were discussed in Chapter 6. Research is needed to identify other ways in which information moves from specialty to specialty within disciplines and from discipline to discipline. Since it appears that scientists require materials outside their own areas, as well as materials directly related to their own areas, further understanding of how they obtain these materials would seem to be imperative if the communication of scientific information is to be improved.

Solving the Communication Problem: Some Recent Proposals
Suggested innovations for dealing with communication problems in science fall into four principal categories: (1) changes in some aspect of the formal communication system, such as the creation of a new type of communications outlet or information service, including replacement of the informal circulation of papers in advance of publication by a formal system that would accomplish the same purpose; (2) improvements in arrangements for oral communciations; (3) replacement of the formal circulation of papers in "packages" in the form of journals by a system of selective dissemination tailored to the needs of the individual scientist; and (4) devices to aid the scientist's personal search of the literature or outright replacement of it by a computerized information retrieval system. The first three types of innovations are primarily designed to increase the "visibility" of materials in the scientist's own research areas. The fourth would aid the scientist in locating materials in other areas.

Circulation of papers in advance of publication appears to be a relatively new innovation among scientists. It seems to have developed spontaneously as a means of providing rapid transmission of information. It is supplemented by informal, face-to-face discussion and by specialized conferences. Attempts are being made to formalize this system (e.g., Dray 1966; Libbey and Zaltman 1967; Van Cott 1970). Circulation of "unrefereed" materials is highly controversial. The principal advantage is that of making available to all scientists in a given area unpublished information that formerly has been available only to the more active scientists and their associates. Those who are relatively isolated from the principal centers of activity in a field would presumably benefit most. A second argument in its favor is that certain types of information such as negative results and results based on small samples that tend to be systematically rejected by the formal communication system would be circulated.

The principal argument against the circulation of un-refereed materials is the necessity for "quality control" (Dray 1966; Etzioni 1971). Herschman (1970) calls the idea of the central manuscript depository "the most serious threat to the institution of journals and to its values." The information science specialist who is primarily concerned with facilitating the transmission of information may be surprised to find that the scientist does not entirely share his concern with speed. The function of the scientific paper that has been refereed and published in a journal is only secondarily to convey information. Its primary function is to serve as a statement of knowledge that has been evaluated and declared acceptable by the scientist's peers (Ziman 1968). Thus "quality control" rather than speed is the most important consideration. This explains a good part of the resistance by scientists to what is probably the largest experiment to date in the circulation of unrefereed materials, the National Institutes of Health information exchange groups, which were eventually withdrawn at the request of the researchers themselves (Dray 1966; Webb 1970). During the period 1961–66, seven information exchange groups were created by the National Institutes of Health. Membership was open to all those who were doing research in these areas. All material submitted by members was duplicated and circulated shortly after receipt. Scientists were concerned that some individuals were obtaining credit for scientific findings on the basis of research that had not been evaluated by knowledgeable peers (Dray 1966). Such systems have other disadvantages. They are expensive. They add to the communication overload by circulating material that may not be of sufficient quality to be published. They may actually reduce the visibility of any particular piece of information in circulation (Loevinger 1970). A recent proposal by the American Psychological Association to circulate to members unrefereed material in their specialties became the subject of considerable controversy (Boffey 1970; Holden 1970). Proponents of the system argued that it is necessary to allow members to cope with the vast amount of material submitted to journals, and that it overcomes the problem of the long lag between submission and publication of manuscripts.

Opinions concerning the necessity for such systems are related to assessments of the performance of the informal system of circulating unpublished materials. Physicists at Stanford who were polled regarding their interest in innovations that would improve their access to information did not appear to be enthusiastic (Parker 1967). Fifty-five percent ($N=75$) failed to respond. Those who did respond claimed that they could obtain by informal means all the information that they needed about their own research area. Significantly, their communication problems occurred when they attempted to obtain information from other fields where they did not have personal contacts. Scientists who are relatively isolated from the informal dissemination of information would probably be more enthusiastic.

The fact that the distribution of productivity in science is highly skewed (a few scientists produce most of the papers, and most scientists produce very few) must also be considered. Griffith and Miller (Project on Scientific Information Exchange in Psychology 1969: 234) found that only 2,200 out of approximately 27,000 psychologists had published at least one paper per year between 1959 and 1963. The author's studies of research areas show that there is a small core of highly productive scientists and a large population of transient and relatively unproductive scientists. The former belong to the informal communication system; the latter often do not. It is uncertain whether the expense and threat to the formal publications system entailed in formalizing the informal dissemination of unrefereed materials is justified in order to enhance the communication of information to these individuals. Parker, Lingwood, and Paisley (1968) have shown that among behavioral science researchers productivity and participation in such a system were correlated (0.31). Their data, however, do not demonstrate a causal relationship. It could be argued that productivity leads to participation in such a system rather than vice versa. If it could be shown that participation in such a system preceded productivity, there would appear to be a stronger case for expanding such systems to include as many persons as possible.

Devices that would permit speedier circulation of articles that have been accepted for publication are not subject to the reservations discussed above. In many fields, the lag between date of acceptance and publication is considerable. Psychological journals have recently been publishing the titles of papers as soon as they are accepted for publication. Garvey and Griffith (1966) studied the characteristics of psychologists who requested advance copies of these papers. "Requesters" were generally young doctorate holders or graduate students. Authors tended to be a somewhat older group. Garvey and Griffith hypothesized that the requesters did not have easy access to the informal system for disseminating information. Their data suggests that this problem is overcome with age and experience in a field, implying that those who would benefit most from the circulation of unpublished materials would also be younger scientists. About a third of the requesters indicated that their work was modified in some way through their contact with the manuscript or its author.

An alternative system (Banks 1970) would be to distribute weekly lists of all papers accepted by all journals in a discipline or possibly in groups of related disciplines. If offered in conjunction with facilities for ordering copies of the manuscripts, this system would speed the dissemination of information. Variations of this approach have been tried by some professional societies (King and Caldwell 1970). The usefulness of most of these systems depends in part upon the amount of agreement regarding the definition of research areas. In disciplines where such consensus is low, mass circulation of lists of papers (either unpublished or in advance of publication) may fail to accomplish their intended purpose. The scientist who is unaware of the labels that others are using to describe the information he needs may fail to spot the relevant titles.

Another innovation, which was suggested over twenty years ago but has not yet been adopted, involves a change in the method of distributing journal articles (Webb 1970). The evaluation process would be maintained, but individual subscribers would receive *only* those articles in which they were interested. The journal itself in the form of issues containing several articles on different topics would cease to appear. This would presumably be more economical than publishing journals in their present form since most journal articles are read by only a small proportion of the subscribers of any journal (Garvey and Griffith 1964). The disadvantages of such a system are primarily two. Again, the categories used to classify articles could be a barrier to wide dissemination of material unless they were continually updated. This would, however, add to the costs of such a venture. The major disadvantage of giving only selected journal articles to any particular journal reader is that it eliminates the browsability factor inherent in journal distribution, making it difficult for the scientist to read outside his field and learn about important innovations of which he ought to be aware. Such a system assumes that the scientist needs only material in his own research area in order to pursue his research effectively. As we have seen, this is not the case.

A number of devices to aid the scientist in his search of the literature already exist. Current-awareness journals that circulate the contents of current periodicals contribute both to the visibility of material and to the browsability of science by making it possible for the scientist to scan a large number of journals without leaving his office. The review article that is becoming increasingly common serves similar functions.

The *Science Citation Index* also provides a way for the scientist to locate information of which he may not have been aware in advance. This index uses a type of subject index that provides one type of solution to the scientist's problem of locating the appropriate classification for the information that he needs (Weinstock et al. 1970). The *Permuterm Subject Index* of the *Science Citation Index* is produced by permuting every significant word in the title of a journal article to produce all possible pairs. This procedure improves the likelihood that the scientist will locate material on his topic when he is unfamiliar with the terminology in current use. Systems that utilize the language that is appearing in scientific titles rather than the standard terms used by indexing services are more responsive to the rapid changes that occur in research-front terminology and are reflected in the article titles rather than in the indexing terms. According to Weinstock and his colleagues, almost two-thirds of the unique or primary words in the *Permuterm Subject Index* of the *Science Citation Index* are new each year.

A computerized information retrieval system would bring "the user of the fund of knowledge into something more nearly like an executive's or commander's position . . . . He will say what operations he wants performed upon what parts of the body of knowledge, he will see whether the results make sense, and then he will decide what to have done next" (Licklider 1965: 32). In an ideal form, such a system is not yet operational. An information retrieval system utilizes either automatic scanning of texts for key words or standard indexing vocabularies in order to identify their contents for the readers. The technique can be applied to the current literature (selective dissemination of information) or used for retrospective searches of the literature.

So far such methods are as effective but no better than manual searches using indexing and abstracting services. A recent review of the literature on this problem concluded that "it is unfortunately the case that all known indexing procedures—whether manual or automatic—produce relatively mediocre results" (Salton 1970: 343). The manual indexing methods are weak because of lack of integration between the classification categories used by the system and those used by the scientists engaged in research. The automatic text analysis procedures yield disappointing results for a different reason. The problem apparently lies in the variety of ways in which terms can be used in scientific documents. Many such uses will be irrelevant to the query of an individual scientist. Analysis of requests for information from 1,000 American physicists showed that the way in which the subject was approached determined the usefulness of information to them. Material on the same subject but approached in a different way was not useful no matter how specifically it was defined (Anthony et al. 1969: 737). This would appear to suggest that more complex procedures involving the use of multiword concepts that reflect document content more fully would be superior, but this proves not to be the case. Salton indicates that these methods may be more effective than single-term indexing for some users but less effective for others. As a result, their over-all performance is low. It is likely that users' queries vary from requests for specific information within a research area to generalized searching for related items that might be relevant. It may be desirable to differentiate between the two types of queries and to use a different type of automatic indexing procedure for each type. An alternative approach that has been used by some researchers in this area is to obtain an evaluation by the user of the materials that have been located midway in the search process and to modify the search procedures accordingly. It is hoped that these "interactive" search methods will improve retrieval effectiveness considerably. It seems likely that further progress in the development of devices of all sorts to speed information-seeking will come about as a result of basic research on how scientists use ideas, which in turn is related to understanding how knowledge develops in an area, since studies of scientists' use of materials that contain ideas, such as books and journals, seem unlikely to yield much new and useful information. For example, studies of changes in the types of materials judged relevant to research as it goes through the various stages of growth would be interesting. Harmon's research on the "cognitive set" suggests that the criteria that scientists use to define and order materials on their subjects become increasingly sophisticated as inquiry proceeds (Harmon 1970).

The review article is a necessary accessory to the increased use of information retrieval, since it makes it possible for the scientist to obtain a broad overview of an area and evaluations of materials in it. Unfortunately, the proportion of review articles published in most disciplines is very small.

Hagstrom (1970) suggests that scientists are not rewarded for writing such articles and that this may explain why they are not produced in greater numbers.

All information retrieval systems are expensive to develop and operate; but, when they have been perfected, they may speed the diffusion of information across research areas and permit the more effective utilization of scientific resources, a problem that is likely to become more serious in the future.

Conclusion

In the preceding pages, it has been argued that the problems of scientific communication can be understood in terms of the interaction between a complex and volatile research front and a stable and much less flexible formal communication system. The research front creates new knowledge; the formal communication system evaluates it and disseminates it beyond the boundaries of the research area that produced it. The fact that the research front is continually evolving and developing makes it difficult for anyone to keep abreast of new findings in a research area solely through the articles appearing in the formal communication system. The research area itself appears to have an effective informal communication network of its own, but not all members of an area participate in it and scientists outside an area are seldom in contact with it. Many of the scientist's difficulties in finding information through the formal communication system stem from the fact that he is often seeking information about areas with which he himself is relatively unfamiliar.

It is clear that the enormous growth of new knowledge is necessitating greater flexibility in the formal communication system. Progress in manipulating this system may come about as a result of increased understanding of the ways in which scientists use ideas and of the types of ideas that are most useful to them. The full range of innovations in the formal communication system has yet to be explored.

8
Toward a Sociology of Culture

In contemporary American sociology, the components of the sociology of knowledge, broadly defined, are studied separately, using a variety of frameworks.[1] A partial list of these components would include the sociology of art, ideology, literature, philosophy, political thought, religion, and science. Little attempt has been made to compare findings from these specialties or to integrate theoretical interpretations. While it is obvious that there are very real differences in the subject matter of each of these areas, stressing their differences and ignoring their similarities have not been fruitful for the development of theory. The sociology of knowledge will only emerge as a useful area of inquiry if a common theoretical model can be developed to explain the whole array of cultural products. If this can be accomplished, it seems likely that the term the sociology of knowledge will cease to be useful. The sociology of culture, being a broader designation, would be more appropriate.[2]

What are the factors that have brought about this unfruitful fragmentation? In part, lack of progress in these areas is

1. For reviews of the area see Adler (1957); Albrecht, Barnett, and Griff (1970); Birnbaum (1960); Curtis and Petras (1970); Fuse (1967); Merton (1957: 456–88, 489–508).
2. Curtis and Petras (1970: 65) mention that Robin M. Williams, Jr., also advocates the use of this term in his lectures to students.

the result of dilemmas in the theoretical model that has been used to study social influences on and consequences of knowledge. First, as was discussed earlier, the preoccupation of traditional theorists, especially Mannheim, with epistemological questions inhibited the development of the field. Second, while Mannheim postulated a relationship between the individual's location in the social structure and the nature of the ideas that he develops and accepts, the nature of the relationship between social structure and cultural products has never been adequately specified (Merton 1957). If social structure influences the development of ideas, how does this influence occur? How does this influence vary with different types of cultural products?

Third, while social class and interest groups have been treated as important influences upon the development and acceptance of ideas, the structure of groups consisting of the producers of ideas has not been discussed (an exception is Coser 1965). The problem of the relationship between the internal structure of a particular cultural institution and the cultural products developed and accepted within it was not explored by the sociology of knowledge. The tendency to view social groups as abstract entities rather than as collections of individuals whose modes of interaction can be precisely observed was probably responsible for this lack. Similarly, the social factors that influence the diffusion of ideas have been only superficially treated in the sociology of knowledge tradition. In order to understand such phenomena, a theory of communications and of the transmission of innovations is needed. This the sociology of knowledge failed to provide.

Sociologists of science and art have begun the preliminary work that must precede serious examination of this problem. For example, sociologists of science have examined the norms, values, reward systems, and communication networks that are peculiar to science (Merton 1957; Hagstrom 1965; Storer 1966). Sociologists of art have begun to specify the relationships between artists, critics, and the public (Duncan 1957). Until recently, however, there has been little attempt to explain the implications of these internal structural characteristics for the development and acceptance of innovations within these institutions.[3]

This task requires analysis of the development of the belief systems of such groups as well as sociometric analysis of the relationships between their members, of the relationships between such groups, and of the relationships of such groups to the larger social structure. In this way, it will be possible to understand the variety of influences that contribute to the development of nonscientific ideas and to their diffusion. In the following pages, it will be shown that seemingly diverse areas, the sciences, the visual and literary arts, and religion, are sufficiently similar in their modes of social and ideational development so that the same theoretical concepts can be applied to all.

Models of Cultural Growth

Price (1963: 69) argues that science is radically different from all other cultural activities. In his view, creative contributions in other cultural areas are uniquely personal. If Michelangelo and Beethoven had never existed, their works would have been replaced by quite different contributions. If Copernicus and Fermi had never existed, however, essentially the same contributions would have been made. There is only one world to discover. The scientist's contributions are not unique to him while those of the artist must be. Price argues that the differences between science and other forms of culture are so basic that any attempt to use similar concepts and models to analyze them would be fruitless.

The impact of Price's argument is diminished by the fact that scientific history has also been written in terms of the contributions of great scientists. Scientific discoveries have been viewed alternatively as the "massive achievements of

3. Selections from the literature on the sociology of knowledge and on the sociology of art and literature recently compiled by Curtis and Petras (1970) and Albrecht, Barnett, and Griff (1970) support the conclusion that specialists in these fields have been predominantly concerned with the relationships between creative individuals and the larger society rather than the relationships among creative individuals.

certain giant figures" (Crutchfield and Krech 1962: 25) and
as the logical outcome of a particular stage in the development of a field. A variation of the latter type of model is as
appropriate for the understanding of art and literature as it
is for science.

Kuhn (1969) admits that it is sometimes difficult to distinguish unambiguously between art and science, and the
problem distresses him. He bases his argument that they are
different in part upon his contention that practitioners of
science and art utilize the histories of their fields in different
ways. He claims that scientists forget the history of their
fields while artists continue to be inspired by theirs. In his
view, new models do not replace old ones in art; rather, the
old ones continue to inspire new developments. If this interpretation were correct, Rembrandt would be as important to
contemporary artists as de Kooning. This model implies that
artistic growth is an almost random process, depending upon
the accidents of individual personalities, entirely unrelated to
the influence of social milieux. Kuhn's view of artistic growth
is similar to Price's description of the growth of literature in
the humanities: unstructured selection from the archives of
the entire field.

It is interesting that an art critic (Gombrich 1966) has
developed a model of artistic growth that is very similar to
the model of scientific growth described by Kuhn (1962)—
cumulative growth punctuated by periodic discontinuities. In
this model, the nature of cultural products is determined by
their creators' perceptions of the nature of reality. Such
"world views" are from time to time readjusted. These re-
adjustments can be drastic, as when an entire theoretical
perspective is altered, or relatively minor, as when a decision
is made to explore a new or previously neglected type of
phenomenon. Kuhn states that "scientific revolutions are here
taken to be those non-cumulative developmental episodes in
which an older paradigm is replaced in whole or in part by
an incompatible new one" (1962: 91).

Gombrich implies that this type of disjunction occurs in
art when he quotes Ernest Jones as having assessed the true

nature of artistic activity as "the constant unmasking of previous symbolisms, the recognition that these, though previously thought to be literally true, were really only aspects or representations of truth, the only ones of which our minds were—for either affective or intellectual reasons—at the time capable" (1963: 30).

Kroeber describes the development of the arts and the development of the sciences in a similar manner. In the arts, also, a particular style is exploited until its possibilities have been exhausted.

But the fine arts . . . come in intermittent spurts instead of going on continuously. It is not too clear what determines where their development stops, except that it seems to be the point at which the values—or ideals, if one likes—that are shaped as the art grows have been attained by the artists' execution, and therewith are exhausted. Unless the goals can now be set farther ahead or widened, and the style be reconstituted on a new basis, there is nothing left for its practitioners but maintenance of the status reached . . . The developmental flow of style is one of its most characteristic qualities (Kroeber 1957: 36–37).

Inherent in these discussions are the three models of cognitive growth that were discussed in Chapter 2. The cumulative view of cultural change in which new ideas develop as logical outcomes of previous ideas has been applied most frequently to science. The second model, which implies that, instead of drawing upon the most recent developments in an area as the inspiration for innovation, new developments can be based upon any previous development, has been considered by some writers to be applicable to nonscientific cultural growth.

The third model is described both in terms of cumulative change with revolutionary discontinuities and in terms of cumulative change until the possibilities inherent in the original ideas have been exhausted. In previous chapters, this model was applied to the development of science. In the following pages, I will discuss the evidence that this model is more useful in understanding the development of nonscientific ideas than the other two models.

Discontinuous Cumulative Growth in Art, Literature, and Religion

The idea of "progress" in the arts is a controversial one. It has had, however, a number of advocates. Gombrich (1966: 9) traces the beginning of progress in the arts to developments that occurred in Florence during the Renaissance. At that point, new pieces of art came to be viewed as "contributions," as "problem-solutions." (The similarity between this concept and Kuhn's idea of a paradigm is startling.) Those who did not internalize the new norm rapidly found themselves left behind by the mainstream of artistic activity. Leadership passed from one major city to another throughout the succeeding centuries, but the idea remained that art was not a series of brilliant and unconnected contributions by great personalities but was a process of discovery in which previous work influenced later contributions.

Herbert Read has said that "for a painter to ignore the discoveries of a Cézanne or a Picasso is equivalent to a scientist ignoring the discoveries of an Einstein or a Freud" (1965). Gombrich has added that "without the idea of one Art progressing through the centuries, there would be no history of art" (1966). As was indicated above, Gombrich does include, however, the possibility of discontinuity in this developmental process. Even a casual survey of the history of art reveals periods when the established view of art has been challenged by relatively marginal artists whose ideas in turn sometimes came to dominate. One thinks of the French Impressionists who rejected the tenets of nineteenth-century representational painting in France, the abstract expressionists who challenged the modern art "establishment" of the 1950s, and the "op art" movement more recently. Each of these groups will be discussed in more detail later in the chapter.

Cumulative growth in science also began during the Renaissance. Science before the Renaissance seems so different to us compared to the science we know now that some scholars have denied that it was really science. The social organization of pre-Renaissance science consisted of masters surrounded by disciples. After the death of the master, the disciples were usually dispersed. As Ben-David (1965) has shown, scientific knowledge was able to evolve and to become cumulative due to the development of appropriate organizational settings that permitted teachers to train students who in turn could assume the role of teacher and continue the earlier work. The organizational factors that permitted art to develop continuously have not yet been studied in a similar fashion. Undoubtedly, the creation of museums, art galleries, and art schools played an important role.

Recent theories of literature have also begun to stress its evolutionary nature. French novelists Nathalie Sarraute (1966) and Alain Robbe-Grillet (1966) have argued that the form of the novel evolves with periodic discontinuities or revolutions. Sometimes a new conception of the novel encounters considerable resistance. Speaking of the power of the nineteenth-century novel in Europe, Sarraute describes how innovations by twentieth-century authors were unable to alter this conception:

The traditional form had such power that it came between the reader and any new work. It was a frame the reader automatically imposed—not only on his own experience but also on all the work with which he came in contact . . . . During these years, unusual work appeared like a freak or a passing trauma. It did not interfere with the peaceful course of the novelistic art. When a novel did not fit the established framework, it became a curiosity, an interesting little monster, an exception to prove the rule (1966: 2).

This description is strikingly parallel to that of Kuhn describing the process of revolution in science: "In the development of any science, the first received paradigm is usually felt to account quite successfully for most of the observations and experiments easily accessible to that science's practitioners . . . . Professionalization leads . . . to an immense restriction of the scientist's vision and to a considerable resistance to paradigm change" (1962: 64).

Art and literature, like science, go through a continual process of evolution and revolution in which old conceptions are replaced by new ones. The violence of the emotions that accompany such shifts is probably related to the degree of commitment to the older conceptions and perhaps to the length of time during which a particular conception has gone unchallenged.

Recently, there has been an attempt to revive evolutionary interpretations of religion. Bellah (1964) has defined a series of stages that characterize this type of phenomenon. Although later stages develop from earlier ones, earlier stages continue to exist alongside later ones. His model is thus more similar to that of random selection from the entire spectrum of previous innovations than to that of discontinuous cumulative development. In this instance, Bellah's emphasis upon the evolutionary aspects of religious thought is important. It suggests the possibility of finding analogies between types of cultural products that have until now been considered too disparate for comparison.

It is not being argued, of course, that the visual arts, literature, religion, and science are identical phenomena. Obviously, there are differences in the nature of these activities, the ways in which materials are treated and analyzed, and in the nature of the conclusions that emerge. The important factor is that science shares with other types of idea systems a similar mode of cumulative development, marked by periodic discontinuities that are sometimes similar to revolutions. This has, as has been suggested, important implications for the nature of their social organization.

The Paradigm, the Social Circle and the Invisible College as Unifying Theoretical Concepts

If the sociology of these various types of idea systems is to be studied using similar frameworks, similar concepts and hypotheses must be applied to all. One such concept is that of the paradigm; others are the social circle and the invisible college.

As we have seen, groups of scientists engaged in developing new research areas are guided by conceptions that are often difficult to define but well understood by those using them. A paradigm indicates what problems need to be solved, what methods should be used to obtain solutions, and what types of phenomena are to be observed. Without such guidelines, a group of scientists would be unable to produce findings that are interrelated and cumulative.

Groups of innovators that produce nonscientific ideas must be guided by similar notions of what is and is not relevant to their interests. Holton (1965) argues that the sharp distinction that is often made between the scientist and the artist is less clear if one examines the innovative process in both areas. In science, this process involves implicit decisions regarding the adoption of certain hypotheses and criteria of selection that are not at all valid in the usual sense of "scientific method." These decisions are influenced by "themata", certain themes that the scientist accepts implicitly and that help "to explain his commitment to some point of view that may in fact run exactly counter to all accepted doctrine and to the clear evidence of the senses." (Holton 1965: 100). They influence what scientists look for, using more precise techniques, and what they do with their findings. Holton claims that a remarkably small number of different themata have played important roles in the development of science. He argues that it is this dimension that science has in common with humanistic scholarship and artistic work.

Groups of artists, like groups of scientists, share orientations toward their work that are not fully or precisely developed. The concept of style that represents the end result of a period of innovation is, as Kuhn has suggested, analogous to theory in the sciences. Artists who are in the process of developing a new style are guided by agreement concerning the importance of particular pieces of art, pictures, or literary creations, which represent the ideas that the group is striving to express. Gombrich (1966) speaks of "problem-solutions" that are embodied in specific art works. Malraux (1965) has stressed that artists learn to paint not from nature but from other paintings.

What has been discussed here is analogous to what was defined earlier as an orientation paradigm. Is there anything in the arts that is comparable to the puzzle-solving devices used by scientists? It is possible that detailed examination of

the ways in which artists use their materials in various me-
diums might reveal something of the sort. One thinks of the
Renaissance artist's discovery of perspective, James Joyce's
experiments with words, Igor Stravinsky's use of bitonality in
his music. Henry Moore recently deplored the problems
facing the contemporary artist who must learn new techniques
and styles that are constantly changing if his work is not to
appear out of date. In this respect, also, the artist is like the
contemporary scientist whose field changes so rapidly that he
must constantly strain to keep abreast of it.

Concepts that have been utilized in the analysis of the
social organization of science should also be useful in understanding other types of cultural phenomena. Kadushin
(1968: 692), in his discussion of the social circle, contends
that there are four kinds of common interests that may provide a focus for such circles, and, thus, there are four kinds
of social circles: cultural, utilitarian, power and influence,
and integrative. Within the first category, cultural, there are
three subcategories: those based on valuational goals such as
religion, psychotherapy, and other philosophies of life; those
based on expressive goals such as literature, art, and recrea-
tion; and those based on cognitive goals such as science and
technology. Examples of utilitarian circles occur in industries
such as "Wall Street," "Seventh Avenue," and "Hollywood,"
whose members need to trade goods and services. Power and
influence circles are found in the areas of community and
national government. Integrative circles are patterns of interaction resulting from some common experience such as
ethnic membership, wartime experience, or membership in an
occupational community.

One way of seeing what the fields that deal with different
aspects of culture have in common is to explore the patterns
of social interaction that exist in such areas. What are the
similarities and dissimilarities among social circles in science,
the visual arts, literature, and other cultural areas?

Analysis of the social organization of research areas in
science has shown that social circles have invisible colleges
that help to unify areas and to provide coherence and direction to their fields. These central figures and some of their

associates are closely linked by direct ties and develop a kind of solidarity that is useful in building morale and maintaining motivation among members. Groups of both these types can be located in the arts and in literature.

For example, White and White (1965) have provided a description of the associations that existed between the leading Impressionists during the years when they were developing their revolutionary approach to painting. They describe how the future Impressionists met each other during the early 1860s and developed close friendships. They worked together in the same small French towns in the provinces, shared apartments, and took painting trips together. As a result, their style during the period when interaction was most intense was a joint creation. The Whites argue that "it was a balance of the contributions of many talents within a given framework. The Impressionists' definition and solution of formal and technical problems was to some degree, then, a result of the social structure of their group and the circumstances of their work in partial isolation from the official system and its styles" (1965: 118).

Here was a small solidarity group surrounded by a looser association of critics, dealers, and buyers who provided recognition, sympathy, and encouragement. Another instance of an artistic community has been well documented in the study by Rosenberg and Fliegel (1965). They concentrated upon examining the social relationships between artists who belonged to the school of abstract expressionism that rose to prominence in New York City after World War II. This group also represented a radical break with the past, and its early experiments provoked "colossal ridicule" from the existing artistic establishment. One of the ways in which they managed to adjust to such a difficult situation was through their sense of solidarity with one another. "It was our community that helped us to break through to a world of marvelous possibilities" (Rosenberg and Fliegel 1965: 34). The authors who examined this group after its high point of community had passed comment: "It was paradoxically the collectivity of artists as a band of brothers, that generated individuality, and when the collectivity collapsed, individuality suffered grievous losses. The unique esthetic vision stands in some sort of organic relationship to a communion, or at least a community of artists" (1965: 36).

Another group of artists which has not been formally studied are the op artists. Like abstract expressionism, this group is centered mainly in a single city, Paris (Musée d'Art Moderne de la Ville de Paris 1967). The pioneers in this area in the early fifties found in Paris an atmosphere of tolerence and sympathy for their efforts and from a few individuals active encouragement. As interest grew, new artists were attracted to this type of art and to Paris as the center of this type of activity. Over one hundred artists were working in this style during the late fifties and early sixties. Although sociometric data is not available concerning their ties with one another, it seems likely that this group resembled a social circle consisting of numerous direct and indirect ties, an invisible college, and several solidarity groups. For example, Le Groupe de Recherche d'Art Visuel consisted of six members who exhibited together and published a joint statement of their artistic views. At least fourteen other groups of "collaborators" could be identified in this area. As scientists in a particular area tend to publish a large proportion of their work in certain core journals, so artists working in the same style tend to exhibit in the same galleries. In this case, fifteen of the forty major artists working in Paris in this style held exhibitions in two galleries.[4]

Such groups also appear in literature and music. Wilson, who has made a thorough study of the role of the poet in American society, has described the relationships between American poets as "a kind of extended family, scattered in space and characterized by both the closeness of interest and the violence of dispute which often mark familial life" (1958: 178). Closer groupings sometimes emerge when new approaches are being developed.[5] These groupings appear to

4. Based on information contained in Musée d'Art Moderne de la Ville de Paris (1967). A "major" artist in this group is defined as one of those selected for inclusion in this major exhibition.
5. For example, the Fugitive Group that was centered at Vanderbilt University in the 1920s. See Cowan (1959).

resemble solidarity groups in that they bolster morale, "inspire a sense of purpose," and provide criticism and "a locus of immediate responsive reward." Older poets sometimes select younger poets for attention and encouragement, lending their names in an honorary capacity to issues of new literary magazines that present the work of their protégés. Group 47, which formed in Germany after the Second World War, also seems to have functioned as a solidarity group, enabling its members to develop a new style in German literature. Finally, Szesztay (1970) has described the solidarity group that surrounded the Hungarian composer Zoltán Kodály and lasted for several decades.

Coser (1965: 3) has argued that all types of intellectuals require regular contact with one another as a result of which they develop common standards for their work and norms for their contact. He contends that most intellectuals cannot produce their work in solitude, that interaction with peers is necessary for the development of ideas. He illustrates his argument with case studies of groups of literary and political intellectuals that have existed during the last three hundred years.

Conclusion

The sociological study of culture must be viewed as a single field using the same concepts and hypotheses to examine different types of cultural phenomena if progress is to be made in understanding this subject. This type of study should be focused upon analysis of the social organization of producers of different types of cultural products and of the themes that guide their creative work. One would expect that social organization in all cultural areas would share certain features such as the maintenance of solidarity and the focusing of interest upon certain types of issues. The questions that should be asked relate to the structure of social organization in different cultural areas and the effect that this structure has upon the development and diffusion of cultural innovations.

Eventually this approach could lead to the systematic analysis of more complex issues such as the interrelationships between different cultural institutions. To what extent are different cultural institutions open to innovations from other such institutions? New movements in one domain are frequently mirrored by new developments in another. In the same way that the numerous and distinct research areas in science are held together by similarities in conceptual orientations and by personal associations, different cultural institutions can also be seen as having similar "world views" during the same historical period and as having interacting memberships. Instead of seeing society as a collection of clearly defined "interest groups," society must be reconceptualized as a complex network of groups of interacting individuals whose membership and communication patterns are seldom confined to one such group alone.

Appendix

Tables

TABLE 1: NUMBERS OF SCIENTISTS LINKED IN COMMUNICATION
NETWORKS OF VARYING DENSITY

Series of Exchanges of Information	Number Each Communicates With			
	3	4	5	6
Starting point	1	1	1	1
First	3	4	5	6
Second	6	12	20	30
Third	12	36	80	150
TOTAL LINKED	22	53	106	187

Note: An exchange of information is defined here as the communication of information by a particular scientist to other scientists whom he knows in the group. When these scientists in turn communicate the information to the scientists they know, this represents a second series of exchanges of information. This model assumes that each scientist receiving information communicates with new scientists and thus that reciprocal and duplicate communications do not occur.

The number of persons with whom the scientist is in communication includes the person from whom he receives information as well as the persons to whom he communicates it.

TABLE 2: GROUP CONNECTIVITY SCORES AMONG SUBGROUPS OF THE RURAL SOCIOLOGY AREA

TYPE OF CHOICE:	SUBGROUP AFFILIATION					Total Group
	High Producers	Moderate Producers	Defectors	Aspirants	Transients	
Informal communication						
Choosing	.205	.237	...	.124	...	.161
Chosen	.468	.283	...	.039	...	.161
Current collaboration						
Choosing	.032	.014	...	.004	...	.010
Chosen	.026	.014	...	.005	...	.010
Published collaboration						
Choosing or chosen	.093	.038	.033	.017	.019	.022
Thesis directors[a]						
Choosing or chosen	.079	.056	.012	.017	.026	.028
Influences on problem selection						
Choosing	.039	.080	.039	.041	.021	.032
Chosen	.214	.023	.127	.007	.015	.032
Influences on technique selection						
Choosing	.018	.030	.001	.013	.005	.009
Choosen	.059	.008	.027	.005	.004	.009

Thesis influence					
Choosing	.005	.010	.001	.005	.003	.004
Chosen	.031	.003	.016	.001	.001	.004
Total ties					
Choosing	.664	.673	.664	.590	.474	.536
Chosen	.801	.518	.802	.392	.542	.536

Note: The group connectivity scores measure the extent to which members of a subgroup are linked to one another and to members of other subgroups. They were computed using figures showing the number of direct and indirect ties for each individual obtained by matrix multiplication of respondents' direct choices performed by the Coleman Sociometric Connectedness Program.

The number of possible ties is obtained by multiplying the number of cases in the group by the number of cases minus one (to eliminate self-choices). The sum of all the individual scores representing the number of direct and indirect relationships with other members gives the number of relationships that actually occur. The group connectivity score represents the proportion of possible ties that actually occur. The number of possible ties for a sub-group is obtained by multiplying the number of cases in the subgroup by the number of cases in the total group minus one. Choosing represents a score based on respondents' own choices. Chosen represents a score based on choices of the respondents by others.

For a description of the subgroups see chap. 3. Only respondents currently engaged in research are included in the first four measures. The remaining measures include all respondents (N=147). Published collaboration also includes nonrespondents (N=221); the nonrespondents' score was 0.015.

[a]These scores are the same for Choosing and Chosen since the relationship between thesis director and student was assumed to be a reciprocal one.

TABLE 3: GROUP CONNECTIVITY SCORES AMONG SUBGROUPS OF THE MATHEMATICS AREA

TYPE OF CHOICE:	SUBGROUP AFFILIATION			Total Group
	High Producers	Moderate Producers	Low Producers	
Informal communication				
Choosing	.078	.055	.088	.078
Chosen	.333	.138	.021	.078
Published collaboration				
Choosing or Chosen	.089	.053	.025	.030
Thesis directors				
Choosing	.012	.015	.008	.010
Chosen	.010	.018	.013	.010
Influences on problem selection				
Choosing	.036	.063	.053	.054
Chosen	.230	.100	.027	.054
Total ties				
Choosing	.405	.463	.354	.379
Chosen	.698	.526	.312	.379

Note: The group connectivity scores measure the extent to which members of a subgroup are linked to one another and to members of other subgroups. They were computed using figures showing the number of direct and indirect ties for each individual obtained by matrix multiplication of respondents' direct choices performed by the Coleman Sociometric Connectedness Program.

For a description of these subgroups see chap. 3. Only respondents currently engaged in research are included in the first two measures. The remaining measures include all respondents (N = 64). Published collaboration also includes nonrespondents (N = 102); the nonrespondents' score was .021.

TABLE 4: DIRECT AND INDIRECT CHOICES RECEIVED BY MEMBERS OF SUBGROUPS IN THE RURAL SOCIOLOGY AREA (in percentages)

	THOSE RECEIVING MORE THAN 10 DIRECT OR INDIRECT CHOICES BY SUBGROUP					
TYPE OF TIE:	High Producers	Moderate Producers	Defectors	Aspirants	Transients	Total Group
Informal communication[a]	100	46	...	9	...	29
Published collaboration	88	27	22	12	15	16
Influences on problem selection	88	9	44	3	3	11
Thesis directors	50	36	0	9	14	16
Total ties[b]	100	64	100	49	68	67
N	(8)	(11)	(9)	(33)	(86)	(147)

Note: The indirect ties between respondents were computed using the Coleman Sociometric Connectedness Program.

[a]Only choices of those members of the research area currently engaged in research in the field are included in these computations.

[b]Also includes current collaboration, thesis influence, and influence on the selection of techniques.

TABLE 5: DIRECT AND INDIRECT CHOICES RECEIVED BY MEMBERS OF SUBGROUPS IN THE MATHEMATICS AREA (in percentages)

| TYPE OF TIE | THOSE RECEIVING MORE THAN 10 DIRECT OR INDIRECT CHOICES BY SUBGROUP | | | |
	High Producers	Moderate Producers	Low Producers	Total
Informal communication	100	39	0	18
Published collaboration	83	40	17	25
Influences on problem selection	75	31	14	16
Thesis directors	0	0	0	0
Total ties	100	77	43	55
N	(4)	(13)	(47)	(64)

Note: The indirect ties between respondents were computed using the Coleman Sociometric Connectedness Program.

TABLE 6: Direct Choices Received by Members of Subgroups of the Rural Sociology Area (in percentages)

Number of Direct Choices Received	Subgroup Affiliation of Scientist Chosen					Total Group
	Transients	Aspirants	Defectors	Moderate Producers	High Producers	
0	43	24	0	0	0	31
1	34	39	11	0	0	29
2–5	16	27	33	73	0	23
6–10	5	9	22	27	0	8
11–20	2	0	22	0	38	5
21–50	0	0	11	0	38	3
Over 50	0	0	0	0	25	1
Total						
Percent	100	99	99	100	101	100
N	(86)	(33)	(9)	(11)	(8)	(147)

Note: Includes respondents only. Percentages in this and subsequent tables do not always total 100 due to rounding error. Choices 1–10 and choices 11 to over 50 were grouped for computation of Goodman and Kruskal's gamma ($\gamma = 0.97$).

TABLE 7: DIRECT CHOICES RECEIVED BY MEMBERS OF
SUBGROUPS OF THE MATHEMATICS AREA
(in percentages)

NUMBER OF DIRECT CHOICES RECEIVED	SUBGROUP AFFILIATION OF SCIENTIST CHOSEN			
	Low Producers	Moderate Producers	High Producers	Total Group
0	42	15	0	34
1	25	5	0	20
2–5	30	35	0	29
6–10	3	40	17	11
11–20	0	5	0	1
Over 20	0	0	83	5
TOTAL				
Percent	100	100	100	100
N	(76)	(20)	(6)	(102)

Note: Includes respondents and nonrespondents.

Choices 1–10 and choices 11 to over 20 grouped for computation of Goodman and Kruskal's gamma ($\gamma = 0.78$).

TABLE 8: GROUPS OF AUTHORS RELATED TO EACH OTHER BY
COLLABORATION AND STUDENT-TEACHER RELATION-
SHIPS IN THE RURAL SOCIOLOGY AREA

Size of Group	Number of Groups	Total Number of Authors in Such Groups	Average Duration of Groups (in years)
1	47	47	1.3
2	9	18	1.1
3	5	15	1.6
4	3	12	5.0
5	1	5	1.0
6	4	24	9.0
7	1	7	7.0
9	1	9	22.0
12	1	12	11.0
13	1	13	12.0
27	1	27	8.0
32	1	32	14.0
TOTAL	75	221	MEAN 2.7

Note: Includes nonrespondents although student-teacher relationships were unknown for them.

Duration of a group of collaborators was measured inclusively by
finding the number of years from the date of the earliest publication
by one of its members to the date of the most recent publication by
a member.

TABLE 9: GROUPS OF AUTHORS RELATED TO EACH OTHER BY COLLABORATION AND STUDENT-TEACHER RELATIONSHIPS IN THE MATHEMATICS AREA

Size of Group	Number of Groups	Total Number of Authors in Such Groups	Average Duration of Groups (in years)
1	41	41	3.6
2	9	18	12.0
19	1	19	41.0
24	1	24	26.0
TOTAL 52		102	MEAN 6.2

Note: Includes nonrespondents although student-teacher relationships were unknown for them.

Duration of a group of collaborators was measured inclusively by finding the number of years from the date of the earliest publication by one of its members to the date of the most recent publication by a member.

TABLE 10: Direct Communication about Current Research by Size of Group of Collaborators in the Rural Sociology Area (in percentages)

| Choices for | Direct Choices Made by Members of | | | |
	Large Groups	Medium Groups	Small Groups and Isolates	Total
Own group	57	32	38	46
Other groups				
Large	21	60	56	40
Medium	17	8	0	11
Small and Isolates	4	0	6	4
Total	99	100	100	101
Number of Choices	(42)	(25)	(16)	(83)
Number of Choices per Member	1.90	1.56	1.14	1.59
Percentage of Members with No Direct Choices	36	44	36	38

Note: The two large groups had 32 and 27 members of which 14 and 8 members, respectively, reported in 1967 that they were continuing to do research in the area. Medium-size groups had between 5 and 13 members of which 2 to 4 were active in each case in 1967, a total of 16 scientists. Small groups had 2 to 4 members of which 1 to 3 were active in each case in 1967, 6 scientists altogether. Eight isolates were active in 1967.

Of the total number of choices, only those of scientists who had continued to do research in the area were included here. A few scientists discussed their research with scientists who were no longer working in the area. These were excluded as were choices of scientists not working in the area.

TABLE 11: INDIRECT LINKS RESULTING FROM COMMUNICATION ABOUT CURRENT RESEARCH BY SIZE OF GROUP OF COLLABORATORS IN THE RURAL SOCIOLOGY AREA (in percentages)

| | INDIRECT LINKS RESULTING FROM CHOICES MADE BY MEMBERS OF | | | |
INDIRECT LINKS WITH	Large Groups	Medium Groups	Small Groups and Isolates	Total
Own group	18	4	5	10
Other groups				
Large	35	60	53	47
Medium	31	20	30	28
Small and Isolates	17	16	12	15
TOTAL				
Percent	101	100	100	100
Number of Indirect Links	(147)	(80)	(105)	(332)
Number of Indirect Links per Member	6.68	5.00	7.50	6.38

Note: The indirect links were computed using the Coleman Sociometric Connectedness Program.

TABLE 12: Direct Communication about Current Research by Size of Group of Collaborators in the Mathematics Area (in percentages)

| CHOICES FOR | DIRECT CHOICES MADE BY MEMBERS OF | | | |
	Large Groups	Pairs	Isolates	Total
Large groups	75	67	65	73
Pairs	4	33	9	6
Isolates	21	0	26	21
TOTAL				
Percent	100	100	100	100
Number of Choices[a]	(72)	(3)	(23)	(98)
Number of Choices per Member[b]	2.88	.60	1.64	2.22
Percentage of Members with No Direct Choices	20	40	50	32

Note: The two large groups had 24 and 19 members of which 16 and 9 members, respectively, reported in 1968 that they were continuing to do research in the area. There were 9 pairs of which five members were still active. Fourteen of the 41 isolates were active.

[a]This total includes choices of active and inactive members and nonrespondents.

[b]Computed for those who were currently conducting research in the area only.

TABLE 13: Indirect Links Resulting from Communication about Current Research by Size of Group of Collaborators in the Mathematics Area (in percentages)

| | Indirect Links Resulting from Choices Made by Members of | | | |
Indirect Links with	Large Groups	Pairs	Isolates	Total
Large Groups	59	64	64	61
Pairs	11	7	9	10
Isolates	30	29	28	29
Total				
Percent	100	100	101	101
Number of				
Indirect Links[a]	(116)	(14)	(47)	(177)
Number of				
Indirect Links				
per Member[b]	4.64	2.80	3.35	4.02

Note: The indirect links were computed using the Coleman Sociometric Connectedness Program.

[a]This total includes choices of active and inactive members and nonrespondents.

[b]Computed for those who were currently conducting research in the area only.

TABLE 14: CHOICES RECEIVED BY MEMBERS OF THE RURAL SOCIOLOGY AREA AND OUTSIDERS (in percentages)

NUMBER OF CHOICES RECEIVED	AFFILIATION OF SCIENTIST CHOSEN	
	Research Area	Outsiders
0	46	...
1	23	63.7
2	9	20.5
3–5	10	12.5
6–10	5	3.0
11–20	4	0.3
21–50	2	0
Over 50	1	0
TOTAL		
Percent	100	100
N	(221)	(389)

Note: Choices were made by members of the research area with respect to informal communication, current collaboration, thesis directors, thesis influence, and influences on the selection of problems and techniques. A scientist could be chosen more than once by the same respondent in different categories.

TABLE 15: CHOICES RECEIVED BY MEMBERS OF THE MATHE-
MATICS AREA AND OUTSIDERS (in percentages)

NUMBER OF CHOICES RECEIVED	AFFILIATION OF SCIENTIST CHOSEN	
	Research Area	Outsiders
0	34	. . .
1	20	75
2	16	19
3–5	14	5
6–10	11	0
11–20	3	0
Over 20	3	0
TOTAL		
Percent	101	99
N	(102)	(93)

Note: Choices were made by members of the research area with respect to informal communication, current collaboration, thesis directors, and influences on the selection of problems. A scientist could be chosen more than once by the same respondent in different categories.

TABLE 16: DISTRIBUTION OF PUBLICATIONS, AUTHORS, AND INNOVATIONS BY FIVE-YEAR INTERVALS IN THE RURAL SOCIOLOGY AREA (in percentages)

	1941–45	1946–50	1951–55	1956–60	1961–66	TOTAL	N
Publications							
Empirical	2	3	12	39	45	101	(329)
Theoretical	0	0	5	30	66	101	(61)
Authors[a]	3	2	13	38	44	100	(203)
Innovations[b]	21	10	20	31	17	99	(201)
Ratio of Innovations to Empirical Publications	7.1	2.3	1.1	.5	.2	.6	
Ratio of Innovations to Authors	7.1	4.2	1.6	.8	.3	.9	

Note: Theoretical publications were not included in the analysis of the diffusion of innovations.

[a] Authors are those publishing for the first time during period indicated. Authors who produced only theoretical publications are not included in this distribution.

[b] Innovations are those appearing for the first time in publications during period indicated.

TABLE 17: INNOVATIONS ADOPTED BY SENIOR AUTHORS BY DATE OF PUBLICATION IN THE RURAL SOCIOLOGY AREA

NUMBER ADOPTERS	DATE OF PUBLICATION					
	1941–45	1946–50	1951–55	1956–60	1961–66	Total
1	16	10	7	16	62	21
2–5	16	24	41	55	35	37
6–20	41	43	49	29	3	32
21–50	11	24	2	0	0	6
Over 50	17	0	0	0	0	3
TOTAL						
Percent	101	101	99	100	100	99
N	(43)	(21)	(41)	(62)	(34)	(201)

$r_{xy} = 0.50$

Note: Each senior author (adopter) is counted once regardless of the number of publications in which he may have used the innovation.

The first adoption represents use of the innovation by the innovator.

In this and tables 18–21, the product-moment correlation coefficient was calculated using ungrouped data.

TABLE 18: Innovations in the Rural Sociology Area
Adopted before and after 1956 (in percentages)

Number of Adopters				
before 1956				
after 1956	1	2–5	6–10	Over 10	Total
Innovations Produced between 1941 and 1955					
0–5	79	28	10	0	49
6–10	10	30	10	0	19
11–20	10	33	10	0	18
Over 20	0	9	70	100	14
Total					
Percent	99	100	100	100	100
N	(48)	(43)	(10)	(4)	(105)
$r_{xy} = 0.89$					

TABLE 19: Adoptions of Innovations Produced at Different Stages of Development of the Rural Sociology Area (in percentages)

Number of Adopters			
1956–60			
1961–66	0–5	6–10	Over 10	Total
Innovations Produced before 1956 ($r_{xy} = 0.90$)				
0–5	82	38	0	66
6–10	17	31	14	18
Over 10	1	31	86	16
Total				
Percent	100	100	100	100
N	(78)	(13)	(14)	(105)
Innovations Produced between 1956 and 1960 ($r_{xy} = 0.07$)				
0–5	93	100	0	94
6–10	5	0	0	5
Over 10	2	0	0	2
Total				
Percent	100	100	0	101
N	(57)	(5)	(0)	(62)

TABLE 20: Innovations Adopted in the Rural Sociology Area by Productivity of Most Productive Adopters (in percentages)

Number of Adopters	Number of Publications by Most Productive Adopter(s)			
	1 to 3	4 to 10	Over 10	Total
3–5	71	67	26	39
6–20	29	30	57	48
Over 20	0	4	17	14
Total				
Percent	100	101	100	101
N	(17)	(24)	(94)	(135)
$r_{xy} = 0.41$				

Note: Since influences upon subsequent adoptions were being assessed, the productivity of last users of innovations was not coded.

This table includes only innovations which were used by three or more scientists including the innovator.

TABLE 21: Innovations in the Rural Sociology Area Adopted by Groups of Collaborators by Date of Innovation (in percentages)

Number of Groups of Collaborators	Date of Innovation		
	1941–55	1956–66	Total
1–2	9	46	24
3–4	30	31	30
5–8	42	23	34
Over 8	19	0	12
Total			
Percent	100	100	100
N	(93)	(65)	(158)
$r_{xy} = 0.64$			

Note: This table excludes innovations which were used only once.

The 203 authors were distributed among 27 groups of collaborators. There were 38 isolates. Isolates were counted as a single group in this analysis.

TABLE 22: Distribution of Publications, Authors, and Innovations by Five-Year Intervals in the Mathematics Area (in percentages)

	Date of Publication						Total	
	1939 and Before	1940–49	1950–54	1955–59	1960–64	1965 to Present	Percent	N
Publications	6	5	9	18	40	22	100	300[a]
Authors	9	8	16	25	26	16	100	101[b]
Publications not citing previous publications	22	15	19	22	13	9	100	55
Proportion of nonciting publications per half-decade	.71	.50	.41	.22	.06	.08	.18	
Ratio of nonciting publications to authors	1.33	1.00	.69	.48	.27	.29	.55	

[a]Does not include six publications that had not appeared at the time the study was done and five publications that could not be obtained.

[b]Excludes one author for whom no publications could be obtained. Includes authors publishing for the first time during period indicated.

TABLE 23: SENIOR AUTHORS CITING AUTHORS IN THE MATHE-
MATICS AREA BY DATE OF AUTHOR'S FIRST PUBLI-
CATION (in percentages)

| NUMBER OF CITERS | AUTHOR'S FIRST PUBLICATION | | | | |
	1939 and before	1940–49	1950–59	1960 to Present	Total Authors
0	0	25	32	66	43
1–10	44	63	46	34	42
Over 10	56	13	22	0	15
TOTAL					
Percent	100	101	100	100	100
N	(9)	(8)	(41)	(44)	(102)
$r_{xy} = 0.29$					

TABLE 24: PUBLICATIONS CITED BY SENIOR AUTHORS AT DIFFERENT STAGES OF THE DEVELOPMENT OF THE MATHEMATICS AREA (in percentages)

NUMBER OF CITERS before 1955 1955 to present	0	1–5	Total
PUBLICATIONS APPEARING BEFORE 1955 ($\phi=0.41$)			
0	52	12	34
1–5[a]	48	69	58
Over 5[a]	0	19	8
TOTAL			
Percent	100	100	100
N	(33)	(26)	(59)
PUBLICATIONS APPEARING BETWEEN 1955 AND 1959 ($\phi=0.26$)			
1955–59 1960 to present			
0	32	6	25
1–5[a]	59	56	58
Over 5[a]	10	38	17
TOTAL			
Percent	101	100	100
N	(41)	(16)	(57)

[a]These categories were grouped for computation of the phi coefficient.

TABLE 25: Publications Cited by Senior Authors in the Mathematics Area by Productivity of Most Productive Citer (in percentages)

| Number of Citers | Number of Publications by Most Productive Citer | | | |
	1 to 3	4 to 10	Over 10	Total
1–5	96	95	59	74
6–10	4	4	32	20
Over 10	0	2	9	6
Total				
Percent	100	101	100	100
N	(23)	(54)	(106)	(183)
$r_{xy} = 0.38$				

Note: 122 publications were not cited.

TABLE 26: PUBLICATIONS CITED BY SENIOR AUTHORS IN THE
MATHEMATICS AREA AT DIFFERENT TIME PERIODS
BY PRODUCTIVITY OF MOST PRODUCTIVE CITER (in
percentages)

NUMBER OF CITERS	NUMBER OF PUBLICATIONS BY MOST PRODUCTIVE CITER		
	10 or Less .	More Than 10	Total
PUBLICATION DATES: BEFORE 1950 ($r_{xy} = 0.57$)			
1–5	100	31	52
6–10	0	58	41
Over 10	0	11	7
TOTAL			
Percent	100	100	100
N	(8)	(19)	(27)
PUBLICATION DATES: 1950–59 ($r_{xy} = 0.31$)			
1–5	88	57	70
6–10	8	34	24
Over 10	4	9	7
TOTAL			
Percent	100	100	101
N	(24)	(35)	(59)
PUBLICATION DATES: 1960–68 ($r_{xy} = 0.36$)			
1–5	98	71	84
6–10	2	21	12
Over 10	0	8	4
TOTAL			
Percent	100	101	100
N	(45)	(52)	(97)

TABLE 27: PUBLICATIONS CITED BY SENIOR AUTHORS IN THE MATHEMATICS AREA AT DIFFERENT TIME PERIODS BY PRODUCTIVITY OF AUTHORS (in percentages)

NUMBER OF CITERS	NUMBER OF PUBLICATIONS BY AUTHORS			
	1–3	4–10	More than 10	Total
PUBLICATION DATES: BEFORE 1950 ($r_{xy} = 0.05$)				
0	12	25	25	18
1–5	59	25	25	42
Over 5	29	50	50	39
TOTAL				
Percent	100	100	100	99
N	(17)	(8)	(8)	(33)
PUBLICATION DATES: 1950–59 ($r_{xy} = 0.25$)				
0	31	44	11	29
1–5	55	41	52	49
Over 5	14	15	37	22
TOTAL				
Percent	100	100	100	100
N	(29)	(27)	(27)	(83)
PUBLICATION DATES: 1960–68 ($r_{xy} = 0.46$)				
0	84	37	30	49
1–5	16	59	49	43
Over 5	0	4	21	8
TOTAL				
Percent	100	100	100	100
N	(57)	(71)	(61)	(189)

TABLE 28: INTERVALS IN WHICH THE NUMBER OF PUBLICA-
TIONS AND NEW AUTHORS FOR SEVEN RESEARCH
AREAS DOUBLED (in years)

RESEARCH AREA	DOUBLING OF PUBLICATIONS	DOUBLING OF NEW AUTHORS
Diffusion of agricultural innovations (rural sociology)	3.0	3.0
Small groups (social psychology)	6.0	—
Mast cells (biomedical research)	8.0	—
Phage (molecular biology)[a]	—	5.5
Theory of finite groups (algebra)	5.0	7.0
Theory of rings (algebra)	3.0	4.5
Theory of determinants and matrices (mathematics)[b]	12.0	—

SOURCES: diffusion of agricultural innovations, Rogers 1966; small groups, McGrath and Altman 1966; mast cells, Selye 1965; phage, Mullins 1968c; theory of finite groups, Gorenstein 1968 and *Mathematical Reviews* 1948–68; theory of rings, Jacobson 1943, 1964; theory of determinants and matrices, Price 1961.

[a]Includes a group of collaborators within the field only rather than all scientists who had published in the area.

[b]A chart appears in Price (1961: 105). This is the only one of the seven fields in which the period of exponential growth occurred during the nineteenth century.

FIGURE 1

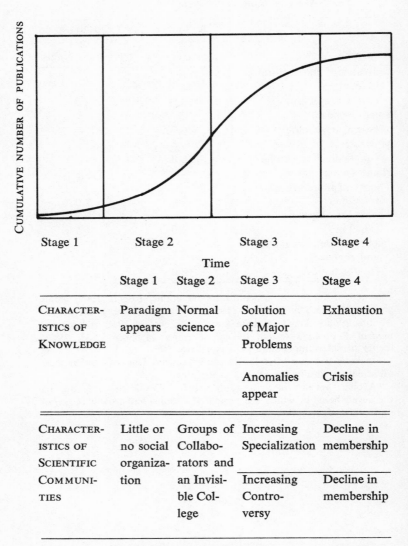

	Stage 1	Stage 2	Stage 3	Stage 4
CHARACTER- ISTICS OF KNOWLEDGE	Paradigm appears	Normal science	Solution of Major Problems	Exhaustion
			Anomalies appear	Crisis
CHARACTER- ISTICS OF SCIENTIFIC COMMUNI- TIES	Little or no social organiza- tion	Groups of Collabo- rators and an Invisi- ble Col- lege	Increasing Specialization	Decline in membership
			Increasing Contro- versy	Decline in membership

Fig. 1. Characteristics of scientific knowledge and of scientific communities at different stages of the logistic curve.

FIGURE 2

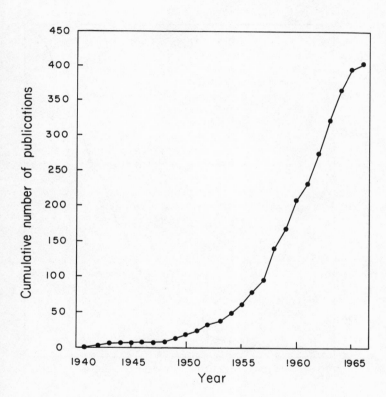

Fig. 2. Cumulative number of publications per year in the rural sociology area (diffusion of agricultural innovations), 1941 to mid-1966. *N* = 403. (Compiled from Rogers 1966.)

FIGURE 3

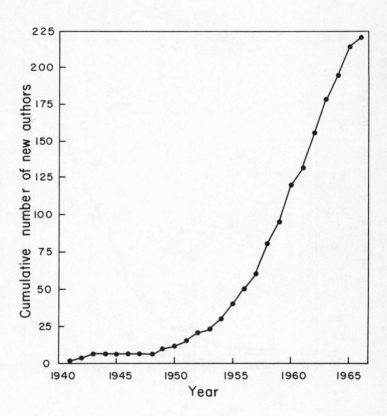

Fig. 3. Cumulative number of new authors per year in the rural sociology area (diffusion of agricultural innovations), 1941 to mid-1966. $N = 221$. (Compiled from Rogers 1966.)

FIGURE 4

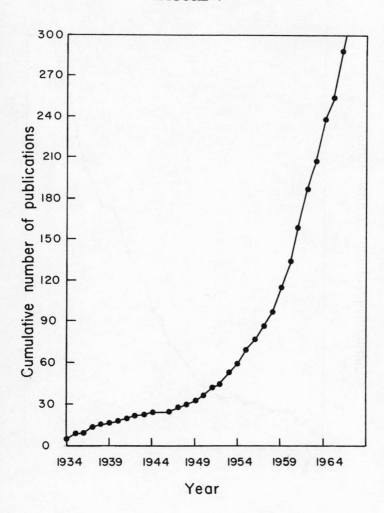

Fig. 4. Cumulative number of publications per year in the mathematics area (theory of finite groups), 1934–68. $N = 305$. (Compiled from Gorenstein 1968 and *Mathematical Reviews* 1948–68.)

FIGURE 5

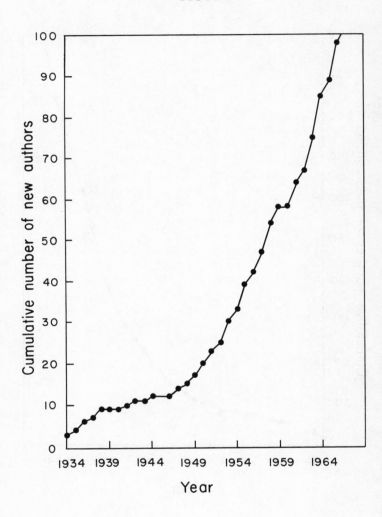

Fig. 5. Cumulative number of new authors per year in the mathematics area (theory of finite groups), 1934–68. $N = 102$. (Compiled from Gorenstein 1968 and *Mathematical Reviews* 1948–68.)

FIGURE 6

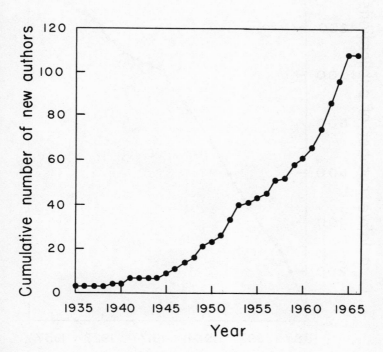

Fig. 6. Cumulative number of new authors per year in the phage area, 1935–66. $N = 108$. (Adapted from Mullins 1968c.)

FIGURE 7

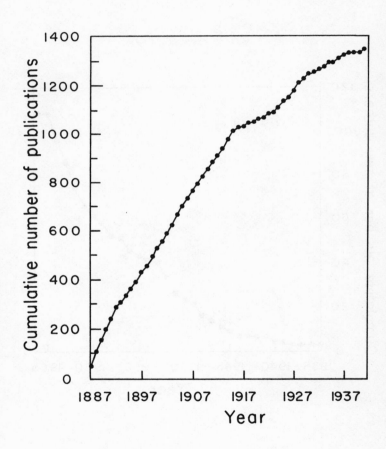

Fig. 7. Cumulative number of publications per year in the invariant theory area, 1887–1941. $N = 1,342$. (Adapted from Fisher 1966.)

FIGURE 8

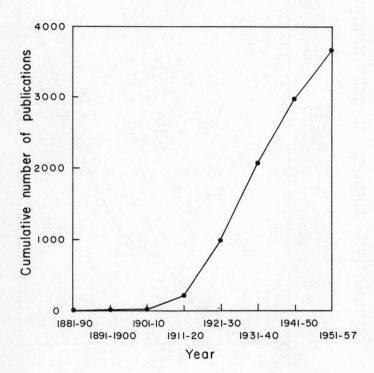

Fig. 8. Cumulative number of publications per year in reading research, 1881–1957. $N = 3{,}684$. (Adapted from Harris 1960, p. 1087.)

FIGURE 9

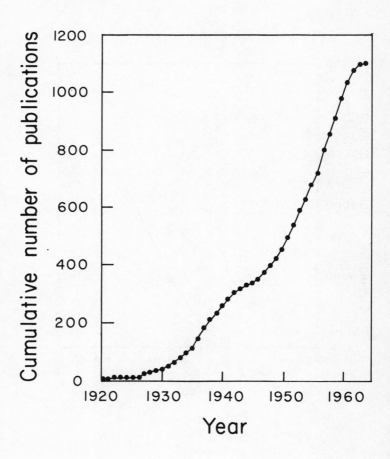

Fig. 9. Cumulative number of publications per year in the theory of rings area, 1920–64. $N = 1,094$. (Compiled from bibliographies appearing in Jacobson 1943 and 1964.)

FIGURE 10

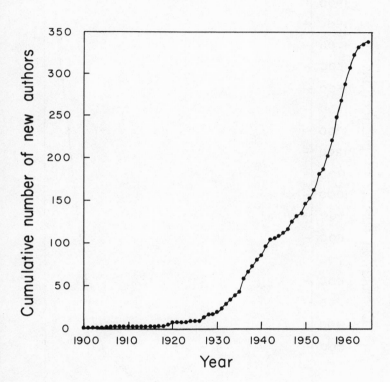

Fig. 10. Cumulative number of new authors per year in the theory of rings area, 1900–1964. $N = 338$. (Compiled from bibliographies appearing in Jacobson 1943 and 1964.)

FIGURE 11

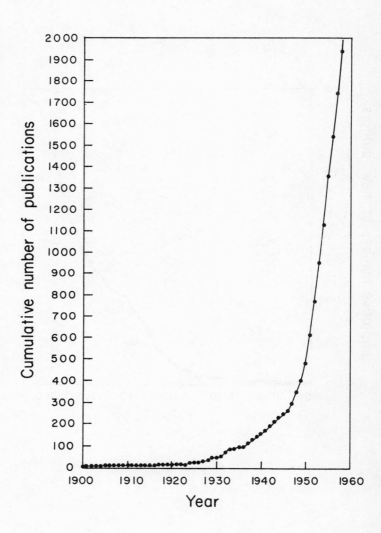

Fig. 11. Cumulative number of publications per year in the small groups area, 1900–1960. $N=2,087$. (Compiled from bibliography appearing in McGrath and Altman 1966.)

FIGURE 12

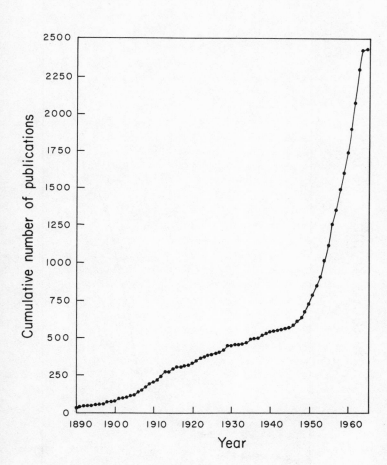

Fig. 12. Cumulative number of publications per year in the mast cells area, 1890–1965. $N = 2,444$ (Compiled from bibliography appearing in Selye 1965.)

FIGURE 13

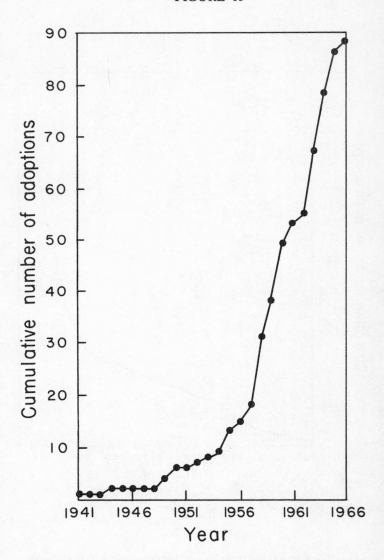

Fig. 13. Cumulative number of adoptions of an innovation by date of adoption in the rural sociology area. $N = 88$.

FIGURE 14

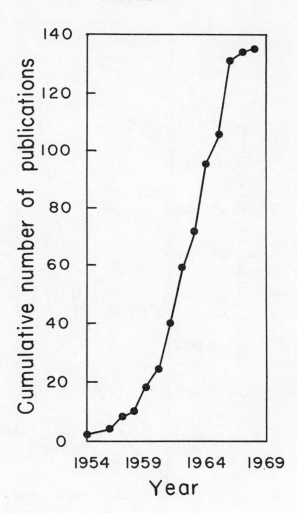

Fig. 14. Cumulative number of publications citing the ten most frequently cited publications in the mathematics area (excluding self-citations and citations appearing in subsequent publications by a previous adopter). $N = 135$.

FIGURE 15

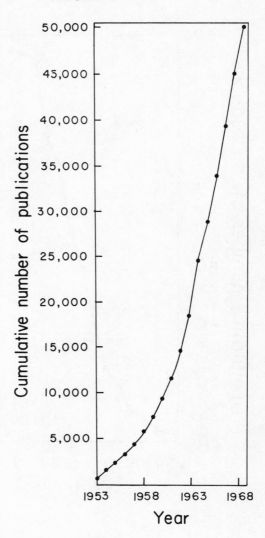

Fig. 15. Cumulative number of publications per year in sociology, 1953–69. N = 51,319. (Compiled from *Sociological Abstracts* 1953–69.)

FIGURE 16

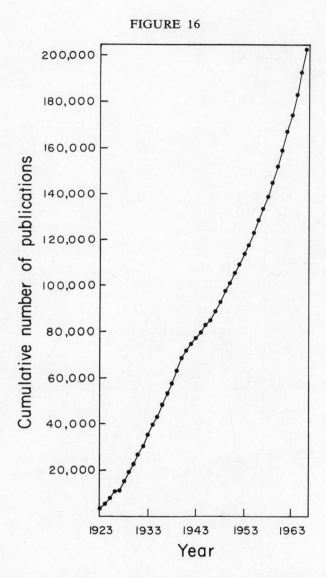

Fig. 16. Cumulative number of publications per year in English language and literature, 1923–67. *N* = 213,899. (Compiled from Modern Humanities Research Association 1923–67.)

FIGURE 17

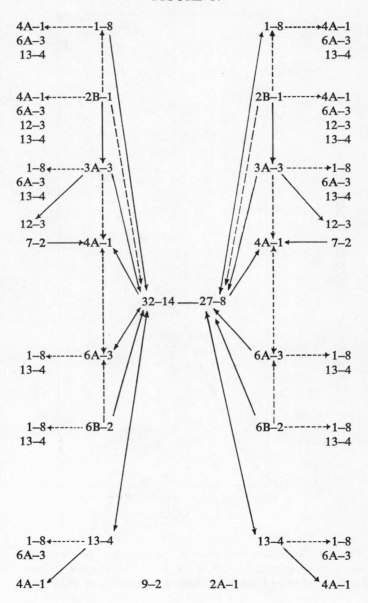

Fig. 17. Direct and indirect communication ties between groups of collaborators in the rural sociology area. The first identification number represents the size of the groups of collaborators including members doing research in the area in 1967 and those who had published in the field but were no longer active in it in 1967. Where more than one group of collaborators had a particular size, letters are used to distinguish between them. The number "1" represents the isolates. The second identification number represents the number of members active in each group in 1967. Only groups in which at least one member was currently doing research are shown.

Direct ties are shown by unbroken lines; indirect ties by broken lines. Arrow signs show the direction of the ties. A group of collaborators is shown as having a tie with another group if any of its members named any member of the other group. An indirect tie between two groups is shown if any member of one group made a choice that led indirectly to any member of the second group.

FIGURE 18

Fig. 18. Direct and indirect influences upon the selection of problems among groups of collaborators in the rural sociology area. The identification number represents the size of the group of collaborators. Where more than one group had a particular size, letters are used to distinguish between them. Direct ties are shown by unbroken lines; indirect ties are shown by broken lines. Arrow signs show the direction of the ties. Only relationships between two medium-size groups that started during the first decade of research in the area and the other groups in the area are shown. Relationships between each of the other groups are not shown. Groups with neither direct nor indirect ties with the two medium-size groups are not shown. Six groups that did not include any respondents were not included in this analysis.

Sample Questionnaire: Mathematics Area

Spring 1968

Yale University
New Haven, Connecticut

SOCIAL ORGANIZATION OF RESEARCH PROBLEM AREAS STUDY

Instructions. Many of the questions can be answered with a check in front of the appropriate answer category. Throughout the questionnaire, guidelines are given in capital letters to summarize the content of each section. When questions can be skipped, the number and page of the next question to be answered is given. In answering sections 1 through 3, please keep in mind the publications by you which are cited in the covering letter. *Any* information you provide will be helpful.

1.1 What is the name of the research specialty and problem area to which the publications cited in the covering letter belong? Please be as specific as possible. For example, my research specialty is the sociology of science; my problem area is the social organization of research specialties. If these publications are related to different problem areas, please name additional areas.
Research specialty _____. Problem area _____.

1.2 Did you write a Master's thesis or a Ph.D. thesis on or related to this problem area(s)?
Master's _____. Ph.D. _____. Neither _____.
IF YES: 1.2.1 Who directed your thesis?
Master's _____. Ph.D. _____.

1.2.2 Have you done any research in this problem area(s) since your thesis? Yes _____. No _____. Please continue with question 2.0.

IF NO: 1.2.3 Please check the statement below that best describes your decision to work in this problem area:
a. Your interest in the area was stimulated *primarily* by publications in the area _____.
b. Your interest in the area was stimulated *primarily* by colleagues _____.
c. Your interest in the area was stimulated *primarily* by activities which you performed as a consultant _____.
d. Other (please specify) _____.

191

2.0 THE FOLLOWING QUESTION CONCERNS THE RESEARCH THAT
YOU HAVE ALREADY DONE ON OR RELATED TO THE PUBLI-
CATIONS CITED IN THE COVERING LETTER.

2.1 List as many persons and/or publications as you wish that
have had an *important* influence upon the *kinds of problems* you have selected for your own research. Please
indicate if you know the authors through their publications
only, through correspondence, or through personal contact
(type of acquaintance). Also please indicate if the person
or publication is concerned with the same problem area(s)
to which the publications cited in the covering letter belong.

Name	Approximate Citation (if individual who influenced you published on the problem)	Type of Acquaintance	Belongs to This Problem Area (Yes or No)

3.0 THE FOLLOWING QUESTIONS CONCERN YOUR CURRENT RE-
SEARCH INTERESTS IN THIS PROBLEM AREA(S).

3.1 Are you *currently* doing research on or related to the research described in the published material cited in the
covering letter? Yes _____. (IF YES: Please continue
with question 3.2 below) No _____. (IF NO: Why have
you not continued to work in this problem area(s)?
_____. [After answering this question please continue
with question 4.])

3.2 Do you discuss your research in this area(s) with anyone?
(Include letters as well as personal contacts.) Yes _____.
No _____. (IF NO: Please continue with question 3.3.)
IF YES: 3.2.1 With whom? Please give names and institutional affiliations and indicate if they are working in this
problem area. Please rate them from 1 to 5 in terms of
the importance of informal communication with them for
your research.

Name	Institutional Affiliation (include researchers at your own university and at other universities, if both are relevant)	Does Research in This Problem Area (Yes or No)	Rating (1: least important; 5: most important)

3.3 Are you presently collaborating with other researchers on
research related to the material cited in the covering letter?
Yes _____. No _____. (IF NO: Please continue with
question 3.4)

IF YES: 3.3.1 With whom? (Please give names and institutional affiliations.)

Name   *Institutional Affiliation*

3.4 Do you have graduate students working in this area(s) at the present time? Yes _____. No _____. (IF NO: Please continue with question 4.0.)
IF YES: 3.4.1 How many? _____.

4.0 THE FOLLOWING QUESTION CONCERNS YOUR RESEARCH INTERESTS IN OTHER AREAS.

4.1 Are you now doing research in other problem areas (either in this research specialty or in another research specialty)? Yes _____. No _____.

5.0 BIOGRAPHICAL INFORMATION.

5.1 Name _____.

5.2 Date of M.S. or M.A. degree _____. 5.2.1 Discipline _____.

5.3 University from which M.S. or M.A. degree was obtained _____.

5.4 Date of Ph.D. _____. 5.4.1 Discipline of Ph.D. _____.

5.5 University from which Ph.D. was obtained _____.

5.6 Professional positions held since receipt of highest degree (indicate name of institution, rank, and period during which each position was held):
Institution _____. Rank _____. Dates _____.

5.7 Have you had research support of any kind to do research on or related to the material cited in the covering letter? Yes _____. No _____.

5.8 On the average, how many *preprints* of publications such as the publications cited in the covering letter do you send?
None _____. 1 to 5 _____. 6 to 10 _____.
11 to 25 _____. 26 to 50 _____. 51 to 100 _____.
101 to 200 _____. More than 200 _____.

5.9 During periods when you are actively working on research related to the material cited in the covering letter, how often do you attend meetings or conferences related to this problem area?
a. More than twice a year _____.
b. Once or twice a year _____.
c. Less than once a year _____.
d. Never _____.

5.10 Have you ever held a visiting appointment at another university or research institute in order to pursue research in this area? Yes _____. No _____.
IF YES: 5.10.1 Where and when? _____.

5.11 Please list up to five journals in which you most frequently find material relevant to your research related to the publications cited in the covering letter. _____.

5.12 Which of the types of positions listed below have you held? (Please check all that apply.)
a. Editor or associate editor of a scientific journal? _____. Which one? _____.
b. President or vice-president of a scientific society? _____. Which one? _____. Title? _____.
c. Member of the executive council of a scientific society? _____. Which one? _____.
Member of a review or advisory committee for a foundation or government agency supporting research? _____. Which one? _____.
None of these _____.

5.13 Have you ever won a prize, special award, or been elected to an honorary scientific society for your research accomplishments? (Do not list prizes or awards received while enrolled in graduate or undergraduate institutions.) Yes _____. Which one(s)? _____. No _____.

5.14 Is the list of your publications in this problem area as cited in the covering letter complete? If not, please list any additional publications of yours in this problem area(s).

Bibliography

Adler, F. 1957. The range of the sociology of knowledge. In H. Becker and A. Boskoff, eds., *Modern sociological theory in continuity and change*, 396–423. New York: Holt, Rinehart and Winston.

Albrecht, M., et al., eds., 1970. *The sociology of art and literature*. New York: Praeger.

Allen, T. J. 1966. The problem solving process in engineering design. *I.R.E.E. Transactions on Engineering Management* 13: 72–83.

_____. 1969a. Roles in technical communication networks. Presented at the Conference on Communication among Scientists and Technologists. Johns Hopkins University, 28 October 1969.

_____. 1969b. Information needs and uses. *Annual Review of Information Science and Technology* 4: 3–29.

Anthony, L. J., et al. 1969. The growth of the literature of physics. *Reports on the Progress of Physics* 32: 709–67.

Back, K. H. 1962. The behavior of scientists: Communication and creativity. *Sociological Inquiry* 32: 82–87.

_____. 1971. *Sensitivity training and the search for salvation*. New York: Russell Sage Foundation.

Baker, D. B. 1970. Communication or chaos? *Science* 169: 739–42.

Banks, E. 1970. Quick publication schemes. *Science* 168: 194–95.

Barber, B. 1961. Resistance by scientists to scientific discovery. *Science* 134: 596–602.

_____. 1962. *Science and the social order.* New York: Collier Books.

_____. 1968. The functions and dysfunctions of "fashion" in science: A case for the study of social change. *Mens en Maatschappij* 43: 501–14.

Barnes, S. B., and Dolby, R. G. A. 1971. The scientific ethos: A deviant viewpoint. *European Journal of Sociology* 11: 3–25.

Barton, A. H., and Wilder, E. D. 1964. Research and practice in the teaching of reading: A progress report. In Matthew B. Miles, ed., *Innovation in education,* 361–98. New York: Bureau of Publications, Teachers College, Columbia University.

Bellah, R. N. 1964. Religious evolution. *American Sociological Review* 29: 358–74.

Ben-David, J. 1960. Roles and innovations in medicine. *American Journal of Sociology* 65: 557–68.

_____. 1964. Scientific growth: A sociological view. *Minerva* 2:455–76.

_____. 1965. The scientific role: Conditions of its establishment in Europe. *Minerva* 4: 15–54.

_____. 1968. National and international scientific communities. Unpublished paper, Hebrew University, Jerusalem, Israel.

Ben-David, J., and Collins, R. 1966. Social factors in the origins of a new science: The case of psychology. *American Sociological Review* 31: 451–65.

Birnbaum, N. 1960. The sociological study of ideology (1940–60): A trend report and bibliography. *Current Sociology* 9: 91–117.

Boffey, P. M. 1970. Psychology: Apprehension over a new communications system. *Science* 167: 1228–30.

Broadus, R. N. 1967. A citation study for sociology. *American Sociologist* 2: 19–20.

Brown, C. H. 1956. *Scientific serials.* Chicago: Association of College and Reference Libraries Monograph Number 16.

Buchanan, J. M. 1966. Economics and its scientific neighbors. In S. R. Krupp, ed., *The structure of economic science,* 166–83. Englewood Cliffs, N. J.: Prentice-Hall.

Burton, R. E., and Kebler, R. W. 1960. The 'half-life' of some scientific and technical literatures. *American Documentation* 11: 18–22.

Cairns, J. et al. 1966. *Phage and the origins of molecular biology*. Cold Spring Harbor, Long Island, N. Y.: Cold Spring Harbor Laboratory of Quantitative Biology.

Campbell, Donald T. 1969. Ethnocentrism of disciplines and the fish-scale model of omniscience. In M. Sherif and C. W. Sherif, eds., *Interdisciplinary relationships in the social sciences*, 328–48. Chicago: Aldine.

Cartter, A. M., ed. 1964. *American universities and colleges*, 9th ed., 1273–76. Washington, D. C.: American Council of Education.

Clarke, B. L. 1964. Multiple authorship trends in scientific papers. *Science* 143: 822–24.

Cole, J. 1970. Patterns of intellectual influence in scientific research. *Sociology of Education* 43: 377–403.

Cole, S. 1970. Professional standing and the reception of scientific discoveries. *American Journal of Sociology* 76: 286–306.

Cole, S., and Cole, J. 1967. Scientific achievement and recognition: A study in the operation of the reward system in science. *American Sociological Review* 32: 377–90.

————. 1968. Visibility and the structural bases of observability in science. *American Sociological Review* 33: 397–413.

Coleman, J. S. 1964. *Introduction to mathematical sociology*. New York: The Free Press of Glencoe.

Coleman, J. S.; Katz, E.; and Menzel, H. 1966. *Medical innovation: A diffusion study*. Indianapolis: Bobbs-Merrill.

Coser, L. 1954. Sects and sectarians. *Dissent* 1: 360–69.

————. 1965. *Men of ideas*. New York: Free Press of Glencoe.

Cowan, L. 1959. *The fugitive group: A literary history*. Baton Rouge, La.: Louisiana State University Press.

Crane, D. 1971. Transnational networks in basic science. *International Organization*, 25 (in press).

Crawford, S. 1970a. Informal communication among scientists in sleep and dream research. Doctoral dissertation. University of Chicago.

————. 1970b. Informal communication among scientists in sleep and dream research. Resumé of dissertation, University of Chicago.

Crews, F. 1970. Anaesthetic criticism. *The New York Review* 14: 31–35.

Crutchfield, R. S., and Krech, D. 1962. Some guides to the understanding of the history of psychology. In L. Postman, ed., *Psychology in the making*, 3–27. New York: Knopf.

Curtis, J. E., and Petras, J. W., eds. 1970. *The sociology of knowledge: A reader*. New York: Praeger.

Dahling, R. L. 1962. Shannon's information theory: The spread of an idea. In *Studies of innovation and of communication to the public*. Stanford, Calif.: Stanford University Institute for Communication Research.

Darwin, C. S. 1958. *Autobiography and selected letters*, ed. Francis Darwin. New York: Dover Publications.

Davis, J. M. 1970. The transmission of information in psychiatry. In American Society for Information Science Annual Meeting, 33d, Philadelphia, 11–15 October 1970. Proceedings, Vol. 7: The Information Conscious Society, 53–56.

De Gré, G. 1970. The sociology of knowledge and the problem of truth. In J. E. Curtis and J. W. Petras, eds., *The sociology of knowledge: A reader*, 661–67, New York: Praeger.

Dessler, A. J. 1970. Swedish iconoclast recognized after many years of rejection and obscurity. *Science* 170: 604–6.

Dodd, S. C. 1955. Diffusion is predictable: Testing probability models for laws of interaction. *American Sociological Review* 20: 392–401.

Dolby, R. G. A. 1971. Sociology of knowledge in natural science. *Science Studies* 1: 1–21.

Dray, S. 1966. Information exchange group no. 5. *Science* 153: 694–95.

Duncan, H. D. 1957. Sociology of art, literature, and music: Social contexts of symbolic experience. In H. Becker and A. Boskoff, eds., *Modern sociological theory in continuity and change*, 482–97. New York: Holt, Rinehart and Winston.

Earle, P., and Vickery, B. C. 1969. Subject relations in science/technology literature. *Aslib Proceedings* 21: 237–43.

East, H., and Weyman, A. 1969. A study in the source literature of plasma physics. *Aslib Proceedings* 21: 160–71.

Etzioni, A. 1971. The need for quality filters in information systems. *Science* 171: 133.

Freeman, H.; Levine, S.; and Reeder, L., eds. 1963 and 1971 (2d ed.). *Handbook of medical sociology*. Englewood Cliffs, N. J.: Prentice-Hall.

Fisher, C. S. 1966. The death of mathematical theory: A study in the sociology of knowledge. *Archives for History of Exact Sciences* 3: 137–59.

————. 1967. The last invariant theorists. *European Journal of Sociology* 8: 216–44.

Fuse, T. 1967. Sociology of knowledge revisited: Some remaining problems and prospects. *Sociological Inquiry* 37: 241–53.

Gamow, G. 1966. *Thirty years that shook physics: The story of quantum theory.* New York: Anchor Books.

Garfield, E., Sher, I. H.; and Torpie, R. J. 1964. *The use of citation data in writing the history of science.* Philadelphia: Institute for Scientific Information.

Garvey, W. D., and Griffith, B. C. 1964. Scientific information exchange in psychology. *Science* 146: 1955–59.

————. 1966. Studies of social innovations in scientific communication in psychology. *American Psychologist* 21: 1019–36.

Gaston, J. 1969. *Big science in Britain: A sociological study of the high energy physics community.* Doctoral dissertation, Yale University.

————. 1970. Rewards, communication, and the division of labor in a scientific community. Paper presented at the Seventh World Congress of Sociology, Varna, Bulgaria, 16 September.

Goffman, W. 1966. Mathematical approach to the spread of scientific ideas—The history of mast cell research. *Nature* 212: 449–52.

Gombrich, E. H. 1963. *Meditations on a hobby horse and other essays on the theory of art.* London: Phaidon Press.

————. 1966. *Norm and form.* London: Phaidon Press.

Gorenstein, D. 1964. Some topics in the theory of finite groups. *Rendiconti di Matematica* 23: 298–315.

————. 1968. *Finite groups.* New York: Harper and Row.

Griffith, B. C., and Miller, A. J. 1970. Networks of informal communication among scientifically productive scientists. In C. Nelson and D. Pollock, eds., *Communication among scientists and engineers,* 125–40. Lexington, Mass.: D. C. Heath.

Hagstrom, W. 1965. *The scientific community.* New York: Basic Books.

_____. 1970. Factors related to the use of different modes of publishing research in four scientific fields. In C. Nelson and D. Pollock, eds., *Communication among scientists and engineers*, 85–124. Lexington, Mass.: D. C. Heath.

Halmos, P. 1957. Nicolas Bourbaki. *Scientific American* 196: 88–99.

Harmon, G. 1970. Information need transformation during inquiry: A reinterpretation of user relevance. In American Society for Information Science Annual Meetings, 33d, Philadelphia, 11–15 October 1970. Proceedings Vol. 7: The Information Conscious Society, 41–43.

Harris, C. W., ed. 1960. *Encyclopedia of educational research*, 3d ed. New York: MacMillan.

Herschman, A. 1970. The primary journal: Past, present, and future. *Journal of Chemical Documentation* 10: 37–42.

Hess, E. L. 1970. Origins of molecular biology. *Science* 168: 664–69.

Holden, C. 1970. APA information plan funded. *Science* 170: 1385.

Holton, G. 1952. *Introduction to concepts and theories in physical science*. Reading, Mass.: Addison-Wesley.

_____. 1962. Scientific research and scholarship: Notes toward the design of proper scales. *Daedalus* 91: 362–99.

_____. 1965. The thematic imagination in science. In G. Holton, ed., *Science and Culture*, 88–108. Boston: Houghton Mifflin.

Hyman, S. E. 1966. *The tangled bank: Darwin, Marx, Frazer, and Freud as imaginative writers*. New York: The Universal Library.

Illinois Institute of Technology Research Institute. 1968. *Technology in retrospect and critical events in science*, Volume 1. Prepared for the National Science Foundation under Contract NSF-C535 (15 December 1968).

Jacobson, N. 1943. *The theory of rings*. New York: American Mathematical Society, Mathematical Surveys, vol. 1.

_____. 1964. *The structure of rings*. American Mathematical Society Colloquium Publication, vol. 37. Providence, R. I.: American Mathematical Society.

Kadushin, C. 1966. The friends and supporters of psychotherapy: On social circles in urban life. *American Sociological Review* 31: 786–802.

————. 1968. Power, influence and social circles: A new methodology for studying opinion makers. *American Sociological Review* 33: 685–99.

Katz, E. 1960. The two-step flow of communication. In Wilbur Schramm, ed., *Mass communications,* 346–65. Urbana: University of Illinois Press.

Katz, E., and Lazarsfeld, P. F. 1955. *Personal influence.* New York: Free Press of Glencoe.

Kaufman, H. F. 1956. Rural sociology 1945–1955. In H. L. Zetterberg, ed., *Sociology in the United States of America,* 104–07. Paris: UNESCO.

Kessler, A. M., and Heart, F. E. 1962. Concerning the probability that a given paper will be cited. Unpublished paper, Cambridge, Massachusetts, Massachusetts Institute of Technology.

Killian, L. M. 1964. Social movements. In R. E. L. Faris, ed., *Handbook of modern sociology,* 426–55. Chicago: Rand McNally.

King, D. W., and Caldwell, N. 1970. Alternatives to the journal system of transferring scientific and technical information. In *Innovations in communications conference,* 9-10 April, 21–35. Springfield, Virginia: Clearinghouse.

King, M. D. 1970. Reason, tradition and the progressiveness of science. Paper read to the British Sociological Association, Sociology of Science Study Group, Durham University, April.

Krantz, D. L. 1965. Research activity in "normal" and "anomalous" areas. *Journal of the History of the Behavioral Sciences* 1: 39–42.

————. 1969. The Baldwin-Titchener controversy. In D. L. Krantz, ed., *Schools of Psychology,* 1–19. New York: Appleton Century Crofts.

————. 1971. Schools and systems: The mutual operation of operant and non-operant psychology as a case study. *Journal of the History of the Behavioral Sciences* 7 (forthcoming).

Kroeber, A. L. 1957. *Style and civilizations.* Ithaca, N. Y.: Cornell University Press.

Kuhn, T. 1962 and 1970 (2d ed.) *The structure of scientific revolutions.* Chicago: University of Chicago Press.

————. 1969. Comment. *Comparative Studies in Society and History* 11: 403–12.

Leach, B. 1971. Scientific techniques in chemistry. Unpublished manuscript, R. & D. Research Unit, Manchester Business School.

Lekachman, R., ed. 1964. *Keynes and the classics.* Boston: Heath.

Libbey, M. A., and Zaltman, G. 1967. *The role and distribution of written informal communication in theoretical high energy physics.* New York: American Institute of Physics.

Licklider, J. C. R. 1965 *Libraries of the future.* Cambridge, Mass.: M. I. T. Press

Lingwood, D. 1968. Interpersonal communication, scientific productivity, and invisible colleges: Studies of two behavioral science research areas. Paper read at the colloquium: Improving the Social and Communication Mechanisms of Educational Research, sponsored by the American Educational Research Association, Washington, D. C., 21–22. November.

Loevinger, J. 1970. Quick publication schemes. *Science* 168: 194.

McGrath, J. E., and Altman, I. 1966. *Small group research: A synthesis and critique of the field.* New York: Holt, Rinehart and Winston.

Malraux, A. 1965. *Museum without walls,* trans. S. Gilbert and F. Price. London: Secker and Warburg.

March, J. G. 1965. Introduction. In J. G. March, ed., *Handbook of organizations, ix-xvi.* Chicago: Rand McNally.

Marquis, D. Q., and Allen, T. J. 1966. Communication patterns in applied technology. *American Psychologist* 21: 1052–60.

Martins, H. 1970. Sociology of knowledge and sociology of science. Unpublished manuscript, University of Essex.

Masterman, M. 1970. The nature of a paradigm. In I. Lakatos and A. Musgrave, eds., *Criticism and the growth of knowledge,* 59–89. Cambridge: At the University Press.

Meadows, A. J. 1967. The citation characteristics of astronomical research literature. *Journal of Documentation* 23: 28–33.

Menzel, H. 1962. Planned and unplanned scientific communication. B. Barber and W. Hirsch, eds., *The sociology of science,* 417–41. New York: The Free Press of Glencoe.

———. 1967. Planning the consequences of unplanned action in scientific communication. In A. de Reuck and J. Knight, eds., *Ciba Foundation symposium on communication in science: Documentation and automation,* 57–71. London: J and A. Churchill.

Merton, R. K. 1957. *Social theory and social structure.* Glencoe, Ill.: The Free Press.

Mikesell, M. W. 1969. The borderlands of geography as a social science. In M. Sherif and C. Sherif, eds., *Interdisciplinary relationships in the social sciences,* 227–48. Chicago: Aldine.

Modern Humanities Research Association. 1923–67. *Annual bibliography of English language and literature.* New York: Cambridge University Press.

Mulkay, M. 1969. Some aspects of cultural growth in the natural sciences. *Social Research* 36: 22–52.

Mulkay, M. 1970. Paradigms and cognitive norms: A working paper. Unpublished manuscript, Aberdeen University.

Mullins, N. C. 1968a. The distribution of social and cultural properties in informal communication networks among biological scientists. *American Sociological Review* 33: 786–97.

―――――. 1968b. Social origins of an invisible college: The Phage group. Paper presented to the American Sociological Association, Boston, August 1968.

―――――. 1968c. The micro social structure of an invisible college: The Phage group. Unpublished manuscript, Dartmouth College, Hanover, New Hampshire.

Musée d'Art Moderne de la Ville de Paris. 1967. *Lumière et mouvement.* Paris: Les Presses Artistiques.

Oberschall, A. 1968. The institutionalization of American sociology. Unpublished paper, Yale University, New Haven Connecticut.

Orr, R. H., and Leeds, A. A. 1964. Biomedical literature: Volume, growth, and other characteristics. *Federation Proceedings* 23, 6, pt. 1 (November-December): 1310–31.

Parker, E. B. 1967. SPIRES (Stanford Physics Information Retrieval System), 1967 annual report. Stanford, Calif.: Stanford University Institute for Communication Research, mimeographed.

Parker, E. B., Paisley, W. J., and Garrett, R. 1967. Bibliographic citations as unobtrusive measures of scientific communication. Stanford, Calif.: Stanford University Institute for Communication Research.

Parker, E. B.; Lingwood, D. A.; and Paisley, W. J. 1968. Communication and research productivity in an interdisciplinary behavioral science research area. Stanford, Calif.: Stanford University Institute for Communication Research, mimeographed.

Polanyi, M. 1962. The republic of science: Its political and economic theory. *Minerva* 1: 54–73.

————. 1967. The growth of science in society. *Minerva* 5: 533–45.

Price, D. J. de S. 1961. *Science since Babylon.* New Haven, Conn.: Yale University Press.

————. 1963. *Little science, big science.* New York: Columbia University Press.

————. 1965a. Networks of scientific papers. *Science* 149: 510–15.

————. 1965b. Is technology historically independent of science?: A study in statistical historiography. *Technology and Culture* 6: 553–68.

————. 1970. Citation measures of hard science, soft science, technology and nonscience. In C. Nelson and D. Pollock, eds., *Communication among scientists and engineers,* 3–22. Lexington, Mass.: D. C. Heath.

Price, D. J. de S., and Beaver, D. 1966. Collaboration in an invisible college. *American Psychologist* 21: 1011–18.

Project on Scientific Information Exchange in Psychology, American Psychological Association. 1969. *Networks of informal communication among scientifically productive psychologists: An exploratory study,* 233–61. Washington, D. C.: American Psychological Association.

Read, H. 1965. *Contemporary British art.* London: Penguin.

Robbe-Grillet, A. 1966. *For a new novel: Essays on fiction,* trans. Richard Howard. New York: Grove Press.

Roberts, A. H. 1970. The system of communication in the language sciences: Present and future. In C. Nelson and D. Pollock, eds., *Communication among scientists and engineers,* 307–23. Lexington, Mass.: D. C. Heath.

Rogers, E. M. 1962. *Diffusion of innovations.* New York: Free Press of Glencoe.

————. 1966. *Bibliography on the diffusion of innovations.* Diffusion of innovations research report, no. 4, Michigan State University, East Lansing.

Rogers, E. M., and Stanfield, J. D. 1966. Adoption and diffusion of new products: Emerging generalizations and hypotheses. Paper presented at the conference on the application of sciences to marketing management, Purdue University, 12–15 July.

Rogers, E. M. et al. 1967. *Codebook for the Michigan State University Diffusion Documents Center.* Working paper 10, Department of Communication, East Lansing, Michigan.

Rosenberg, B., and Fliegel, N. E. 1965. *The vanguard artist.* Chicago: Quadrangle Books.

Russett, B. R. 1968. Methodological and theoretical schools in international relations. Unpublished paper, Yale University, New Haven, Connecticut.

Salton, G. 1970. Automatic text analysis. *Science* 168: 335–43.

Sarraute, N. 1966. The novel for its own sake. *The New York Times Book Review,* 24 April, 2, 43.

Selye, H. 1965. *The mast cells.* Washington, D. C.: Butterworths.

Slater, M., and Keenan, S. 1967–68. Current papers in physics study: reports 1-3. London: Institution of Electrical Engineers; New York: American Institute of Physics.

Stevens, R. E. 1953. Characteristics of subject literatures. *ACRL Monographs,* no. 6, 10–21.

Stoddart, D. R. 1967. Growth and structure of geography. *Transactions and papers of the Institute of British Geographers,* publication no. 46.

Stone, R. 1966. *Mathematics in the social sciences and other essays.* Cambridge, Mass.: M.I.T. Press.

Storer, N. 1966. *The social system of science.* New York: Holt, Rinehart and Winston.

————. 1968. Modes and processes of communication among scientists: Theoretical issues and prospects for investigation. Paper presented at Conference on Theoretical Issues in the Study of Science, Scientists, and Science Policy, sponsored by the Social Science Research Council and the Institute for the Study of Science in Human Affairs, New York, 29 February to 1 March 1968.

Swanson, D. R. 1966. Scientific journals and information services of the future. *American Psychologist* 21: 1005–10.

Szesztay, A. 1970. Scientific "schools" in Hungary: Zoltán Kodály and his disciples. Paper presented at the Seventh World Congress of Sociology, Varna, Bulgaria, 16 September 1970.

Toulmin, S. 1963. *Foresight and understanding.* New York: Harper and Row (Torchbook Series).

————. 1966. Is there a limit to scientific growth? *Science Journal* 2: 80–85.

————. 1967. Conceptual revolutions in science. In R. S. Cohen and M. W. Wartofsky, eds., *Boston studies in the philosophy of science.* Volume 3, 331–47. Dordrecht, Holland: D. Reidel Publishing Company.

————. 1971. New directions in philosophy of science. *Encounter* 35: 53–64.

Van Cott, H. P. 1970. National information system for psychology: A proposed solution for a pressing problem. *American Psychologist* 25: i–xx.

Walum, L. 1963. Pure mathematics: A study in the sociology of knowledge. Doctoral dissertation, University of Chicago.

Webb, E. C. 1970. Communication in biochemistry. *Nature* 225: 132–35.

Weinstein, N. n.d. The development of theoretical cosmology, 1963–65; or "Que sçais-je?" Unpublished manuscript.

Weinstock, M. F. et al. 1970. System design implications of the title words of scientific journal articles in the *Permuterm Subject Index:* I. Conformity to Zipf's Law. Presented at the 7th Annual National Information Retrieval Colloquium, Philadelphia, Pennsylvania, 2–8 May.

White, H., and White, C. 1965. *Canvases and careers.* New York: Wiley.

Willer, D., and Webster, M. 1970. Theoretical concepts and observables. *American Sociological Review* 35: 748–57.

Wilson, R. N. 1958. *Man made plain.* Cleveland: Howard Allen.

Xhignesse, L. V., and Osgood, C. E. 1967. Bibliographical citation characteristics of the psychological journal network in 1950 and 1960. *American Psychologist* 22: 778–91.

Yinger, J. M. 1957. *Religion, society and the individual.* New York: MacMillan.

Zaltman, G. 1968. *Scientific recognition and communication behavior in high energy physics.* New York: American Institute of Physics.

Zaltman, G., and Blau, J. 1969. A note on an international invisible college in theoretical high energy physics. Unpublished paper, Northwestern University.

Zaltman, G., and Köhler, B. M. 1970. Transaction flows and diffusion of research specialties in an international scientific community. Unpublished paper, Northwestern University.

Ziman, J. M. 1968. *Public knowledge: The social dimension of science.* Cambridge: Cambridge University Press.

Zuckerman, H. 1967. Nobel laureates in science: Patterns of productivity, collaboration, and authorship. *American Sociological Review* 32: 391–403.

Index

TR notes

p 1,
p 12
p 16: single - author paper

pp 24-5
p 41
p 47 (but every weasel communal
in Britain)
p 4 9, indirect lineage,
123)
) 53
62: preprints
65: phys in. coll + Japan